selected mathematical derivations for engineers

allan de medeiros martins

27/07/2014

First Edition

Federal University of Rio Grande do Norte (UFRN)
Produced at UFRN @ Department of Electrical Engineering (Technology Center)

ISBN 978-1-312-39053-9

To anyone that loves Science...

... and to my mom and dad that love it just because it makes me happy!

Special thanks to my beautiful and beloved wife Ana Beatriz♡ for the love, patience, comprehension and strength ...

Contents

Preface

This book, as the title suggests, is a collection of mathematics derivations which normally are considered difficult for students to grasp. The goal is to try to give detailed derivations of formulae, concepts or theorems in order to facilitate the understanding of those subjects. We will try to balance the formality of the mathematics behind the derivations with some informal definitions and statements. This book is not intended for mathematics students who seek a completely formal derivation of a specific topic. Instead, it is directed to students (engineering, mathematics and science in general) who seek to understand specific concepts used in areas like engineering, mathematics, statistics etc. It follows a somewhat formal format in order to be able to go deeper to a more formal version of those derivations. As the student understands the concepts behind the derivation, we hope that he/she could look for a more in-depth version and be able to completely understand it.

The book is a collection of unrelated topics. There is no direct relationship between the subjects and they are not connected in a specific way. The content of the book is the result of some assorted classes that the author gave during his teaching career. It was motivated by comments and criticisms by very good students who felt that somehow those classes helped them understand concepts that they found difficult at first glance.

Any chapter can be read independently. There is no order among the chapters, although in each chapter, there are topics that can be regarded as having some order (for instance, in the

Fourier transform section, it is better to read the content in sequence). The book is to be used as a reference for studies which require the understanding of the mathematics behind the subjects. Although there are not many topics, one can look at it as a dictionary of "derivations". Because the book's purpose is to present mathematical derivations, there will be a lot of mathematical notation and the derivations will be as detailed as is necessary for the student to fully follow the steps. Sometimes that means the topics will be practically a "follow the maths" exercise, and there will be as many topics as possible. Hopefully, in future editions, more and more topics will be added. Again, there is no logic or connection among the topics ... they are only a collection of classes from one author, nothing more.

It is impossible to mention everyone who helped me in some way with this book. There were so many people who directly or indirectly contributed or motivated me and I'm thankful to all of them. I won't take the risk of forgetting to mention anyone, so thank you all, my friends, family, students, professors, colleagues and co-workers.

Finally, if anyone has any comment suggestion or criticism, feel free to contact the author at (currently) `allan@dee.ufrn.br` or in the social media available at the moment. Also, you are welcome to try the author's apps: the game "can Ŭ control" and the dynamical systems simulation app "iDynamic". Both apps explores mathematical concepts present in this book and are available (currently) at `http://www.ngmsoftware.net/canucontrol` and `http://www.ngmsoftware.net/idynamic` respectively.

Chapter 1

Introduction

Nobody doubt that in science and engineering, mathematics is the most important tool available. As a way to visualise the role of mathematics in science, it is useful to think of a hierarchy of areas. Each area is used to understand the world and at the very bottom is physics. Physics explains the atoms, the forces and the fields that there are in nature. As the number of particles increases, molecules arise and physics loses its accuracy because the systems become too complex to be analysed with the laws of physics. Then, one makes some assumptions and simplifications and we have chemistry, which explains substances, polymers, and small clusters of molecules. But again, as the molecules clump together, the large objects that are formed become too complex, exhibiting autonomous behaviours and so are again too complex to be analysed. Now, simplifications and assumptions give rise to the field of biology which studies living things and complex molecules like proteins, fats, DNA, etc. This path branches into the areas of medicine, engineering, social sciences etc. One must now ask where does mathematics come in? The most suitable answer is that mathematics is present in every step of that path as the language used to express and communicate all that knowledge in an unambiguous way.

Mathematics is used more or less in the same way in all

these areas. One makes statements, writes them down using symbols and rules and derives conclusions in the form of statements or formulae. For example, everybody knows how to compute the average or mean of a set of numbers. We just sum up the numbers and divide by the quantity. But, why is the average computed that way? Another more specific example is about the controllability matrix for linear systems.Engineering students know that to test for controllability one must take the matrix $\mathbf{R} = [\mathbf{B} \quad \mathbf{A}\mathbf{B} \quad \mathbf{A}^2\mathbf{B} \quad \cdots]$ and test for $rank(\mathbf{R}) = n$. In practice, not all students know why they must do that, although they know how to do it. This book aims to help students to know *why* they do those things by presenting the derivations behind those formulae and statements.

The derivations presented here are not specifically related in any way. They are just derivations that are either difficult for the student to do alone or have an interesting interpretation. The subjects are divided by disciplines, each one composing a chapter. In each chapter, assorted derivations are presented as sections. The fact that the book presents several different subjects, imposes the need for some prior knowledge. In each section, the focus is on presenting the derivation of the formula or statement alone. Practically no information about the problem itself is given or detailed. For instance, in the section about observability and controllability, the reader will not find much explanation of what observability or controllability is. Therefore, it is required that the reader be familiar with some basic methods and theories (for instance that to optimise some function, one must differentiate and equate to zero).

Throughout the book, the notation will be somewhat consistent. It is very difficult to keep it completely consistent since we have different areas that use different letters for common quantities (like $\mathbf{R}$ for the reachability matrix in control theory and the auto-correlation matrix in signal processing). We will use bold lowercase for vectors and bold uppercase for matrices. Specific notations will be explained in the text as they are needed (as L for operator instead of a variable). Whenever we reach a proof of some statement or reach a final form for a formula we will end the preceding sentence with a "black square" symbol ■.

The chapter two is dedicated to calculus. There is a large subdivision into the topics of differentiation, integration and some numerical methods. In the chapter three there is some topics in Probability and Statistics and no special subdivision is present. In chapter four the presented topic is Signal Processing. Again no subdivision is present, although the subsection of Fourier analysis is subdivided into the for categories depending on the class of the signal. Chapter five is the larger one. The topic for that chapter is Control Theory and the chapter is basically divided into continuous and discrete dynamical system analysis. Chapter six shows some derivations about Machine Learning and has no special subdivision either. In chapter seven we present some assorted derivations in miscellaneous topics. Those topics have no particular link and some are just fun derivations (like prooving that a line segment can have any length).

Chapter 2

Calculus

In this chapter we will show the derivations for some common formulae in calculus. These are very useful in proving the next chapter's formulae. Nevertheless, understanding the steps in these proofs hopefully will give the reader insights and tips on how to perform their own proofs for other theorems or formulae that might be of interest. This is not even close to a review or a text about calculus, but rather, it is (as the book is entitled) a collection of selected derivations that might help the advanced student. As the book also says, the derivations are not as formal as required for a mathematics graduate text. The idea is to provide a middle ground between illustrative and formal understanding of the concepts. For instance, we might use terms like "very small" to mean "differential" and will not bother to prove that Δt^2 is zero in the limit when $\Delta t \to 0$. Rather we will use textual explanations such as "Δt^2 reaches zero first than Δt" and so on.

2.1 L'Hôpital's rule

L'Hôpital's Rule is a very useful tool for computing limits. It states that the limit of a quotient of two functions is equal to the limit of the quotient of its derivatives. Formally we can write

$$\lim_{x \to x_0} \frac{f(x)}{g(x)} = \lim_{x \to x_0} \frac{\dot{f}(x)}{\dot{g}(x)} \tag{2.1}$$

The problem with this kind of limit arises when the two functions goes both to zero or infinity, leading to a mathematical indetermination. In those cases, the derivative of one or both might not go to zero or infinity as well and the limit can be computed. Next, we will demonstrate why that is true.

2.1.1   Case 0/0

We can start by plugging in the definition of the derivative for both functions, without making the limit. By doing that we obtain

$$\frac{f(x + \delta) - f(x)}{\delta} = \dot{f}(\xi_1)$$
$$\frac{g(x + \delta) - g(x)}{\delta} = \dot{g}(\xi_2) \tag{2.2}$$

where $x = x_0$ is the point of the limit. ξ_1 and ξ_2 are values in the interval

$$x < \xi_1 < x + \delta$$
$$x < \xi_2 < x + \delta \tag{2.3}$$

and the equality always holds for continuous functions (which is the condition for the L'Hôpital's Rule to work). That is guaranteed by the Mean Value Theorem.

Now, we make the quotient

$$\frac{\frac{f(x+\delta)-f(x)}{\delta}}{\frac{g(x+\delta)-g(x)}{\delta}} = \frac{\dot{f}(\xi_1)}{\dot{g}(\xi_2)} \tag{2.4}$$

and simplify it to

$$\frac{f(x + \delta) - f(x)}{g(x + \delta) - g(x)} = \frac{\dot{f}(\xi_1)}{\dot{g}(\xi_2)} \tag{2.5}$$

If both $f(x)$ and $g(x)$ are going to zero, the limit when $\delta \to 0$ becomes (notice that by 2.3, when δ goes to zero, both ξ_1 and ξ_2 goes to x)

$$\frac{f(x+\delta)}{g(x+\delta)} = \frac{\dot{f}(\xi_1)}{\dot{g}(\xi_2)}$$

$$\lim_{x \to x_0} \frac{f(x)}{g(x)} = \lim_{x \to x_0} \frac{\dot{f}(x)}{\dot{g}(x)}$$

(2.6)

as L'Hôpital's Rule states ■.

2.2 Differentiation

In this section we will try to illustrate basic definitions for differentiation of functions in several cases (one variable, many variables etc.). The goal is not to "demonstrate" formally any concept, but rather to give examples and illustrations of the meaning of each case of differentiation in Euclidian space. We hope that the reader grasps the idea and the difference between objects like $\frac{d}{dt}$ and $\frac{\partial}{\partial t}$ (total and partial derivatives) since it is a source of much confusion among engineering students.

2.2.1 One function of one variable

The case where we have one function of one variable is the simplest case of differentiation. We simply have some value as a function of another $y = f(x)$. In this case the definition for the differentiation is

$$\frac{\mathrm{d}f(x)}{\mathrm{d}x} = \lim_{\Delta x \to 0} \frac{f(x + \Delta x) - f(x)}{\Delta x}$$

(2.7)

Graphically, we have it representing the tangent of the function at the point x as shown in Figure 2.1

When we have a composite function $f(x(t))$, the differentiation of f with respect to t can be obtained by the Chain Rule as

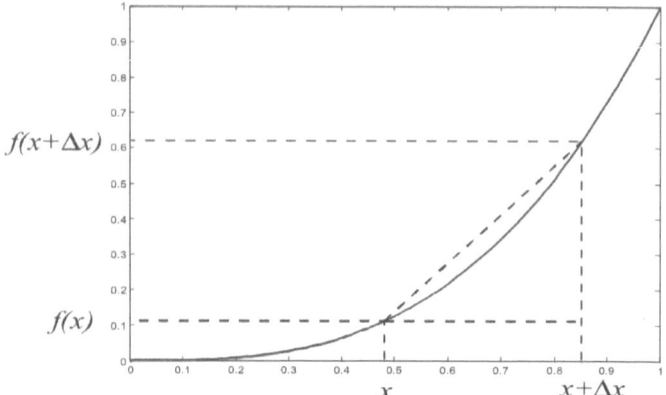

Figure 2.1: Line representing the definition for the derivative of a function. In the limit $\Delta t \to 0$ the line will be tangent at the point x.

$$\frac{\mathrm{d}f(x)}{\mathrm{d}t} \approx \frac{\Delta f(x)}{\Delta t}$$
$$\approx \frac{\Delta f}{\Delta t}\frac{\Delta x}{\Delta x} = \frac{\Delta f}{\Delta x}\frac{\Delta x}{\Delta t} \qquad (2.8)$$

Taking the limit for x and t we have

$$\frac{\mathrm{d}f(x)}{\mathrm{d}t} = \frac{\mathrm{d}f(x)}{\mathrm{d}x}\frac{\mathrm{d}x(t)}{\mathrm{d}t} \qquad (2.9)$$

Product rule

We are interested in developing the expression for the derivative of the product

$$\frac{\mathrm{d}(u(t)v(t))}{\mathrm{d}t} \qquad (2.10)$$

Let us write approximations for the functions $v(t)$ and $u(t)$ using individual derivatives as

$$u(t + \Delta t) \approx u(t) + \frac{du(t)}{dt}\Delta t$$

$$v(t + \Delta t) \approx v(t) + \frac{dv(t)}{dt}\Delta t$$

(2.11)

Now, we use the definition for the derivative in the product

$$\frac{d(u(t)v(t))}{dt} \approx \frac{u(t + \Delta t)v(t + \Delta t) - u(t)v(t)}{\Delta t} \qquad (2.12)$$

We can use 2.11 to find out the product $u(t + \Delta t)v(t + \Delta t)$ as

$$u(t + \Delta t)v(t + \Delta t) = \left(u(t) + \frac{du(t)}{dt}\Delta t\right)\left(v(t) + \frac{dv(t)}{dt}\Delta t\right)$$

$$= u(t)v(t) + v(t)\frac{du(t)}{dt}\Delta t + u(t)\frac{dv(t)}{dt}\Delta t + \frac{du(t)}{dt}\frac{dv(t)}{dt}\Delta t^2$$

(2.13)

Now, we can substitute 2.13 in 2.12 and get (neglecting the term with Δt^2 as it tends to zero faster than the others)

$$\frac{d(u(t)v(t))}{dt} \approx \frac{u(t)v(t) + v(t)\frac{du(t)}{dt}\Delta t + u(t)\frac{dv(t)}{dt}\Delta t - u(t)v(t)}{\Delta t}$$

$$= \frac{v(t)\frac{du(t)}{dt}\Delta t + u(t)\frac{dv(t)}{dt}\Delta t}{\Delta t}$$

(2.14)

which (in the limit) results in the famous Product Rule ∎

$$\frac{d(u(t)v(t))}{dt} = v(t)\frac{du(t)}{dt} + u(t)\frac{dv(t)}{dt} \qquad (2.15)$$

2.2.2 One function of several variables

The case when we have one function of several variables is a bit more elaborated. Now we have some value as a function of two or more values (for example $z = f(x,y)$). In this case the definition for the differentiation is

$$\frac{\partial f(x_1,\ldots,x_n)}{\partial x_1} = \lim_{\Delta x_1 \to 0} \frac{f(x_1 + \Delta x_1, x_2, \ldots, x_n) - f(x_1, x_2, \ldots, x_n)}{\Delta x_1}$$

(2.16)

which can be defined for each variable. Notice that now the symbol ∂ do not always mean "increment" because ∂f can be an increment in any direction. Rather, the whole symbol $\frac{\partial f}{\partial x_i}$ represents the slope of $f(x_1, \ldots, x_n)$ in the direction of x_i (keeping all the other variables constant).

Graphically, that slope is measured at the point $(x_1, x_2, \ldots, x_n)$ as shown in Figure 2.2

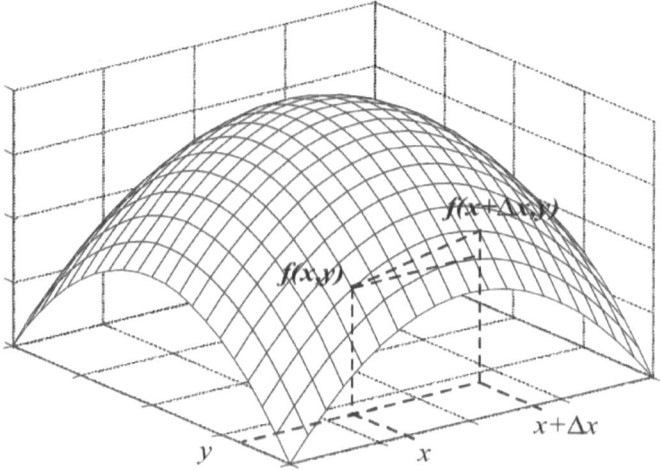

Figure 2.2: Line representing the definition for the derivative of a function of several variables. In the limit $\Delta x \to 0$ the line will be tangent at the point (x,y) in the x direction.

When we have a composite function $f(x(\theta_1, \theta_2), y(\theta_1, \theta_2))$ the differentiation of f with respect to θ_1 or θ_2 can be obtained by the chain rule. To obtain a chain rule for this case we can't simply multiply by $\frac{\partial \theta_1}{\partial \theta_1}$ because the partial's symbols can't be broken and swapped. That is not possible because the partial is only a fraction when you keep all the other variables constant. Since x and y depend on θ_1, both will change and the symbol $\frac{\partial f}{\partial \theta_1}$ cannot be considered a fraction. The solution is to write the increment in f due to both changes as

$$\Delta f = \frac{\partial f(x,y)}{\partial x} \Delta x + \frac{\partial f(x,y)}{\partial y} \Delta y \qquad (2.17)$$

Now, we divide both sides by $\Delta \theta_1$ and obtain

$$\frac{\Delta f}{\Delta \theta_1} = \frac{\partial f(x,y)}{\partial x} \frac{\Delta x}{\Delta \theta_1} + \frac{\partial f(x,y)}{\partial y} \frac{\Delta y}{\Delta \theta_1} \qquad (2.18)$$

Since we are keeping θ_2 constant all the time, all the fractions represent slopes in θ_1 and therefore can be written as partials. So, in the limit we have ■

$$\frac{\partial f}{\partial \theta_1} = \frac{\partial f(x,y)}{\partial x} \frac{\partial x(\theta_1, \theta_2)}{\partial \theta_1} + \frac{\partial f(x,y)}{\partial y} \frac{\partial y(\theta_1, \theta_2)}{\partial \theta_1} \qquad (2.19)$$

2.2.3 Several functions of one variable

In this case, we have only one variable (say t) and several functions represented as a vector that depend on the variable. Those functions are sometimes called parametric functions. The notation used is the following where $v_i(t)$ are the functions, and n is the number of functions or dimension of the vector. We use lowercase bold for vector variables

$$\mathbf{v}(t) = [v_1(t) \quad v_2(t) \quad \ldots \quad v_n(t)]^t \qquad (2.20)$$

The differentiation in this case is done for each function and is performed in the variable t. This results in another vector, defined as

$$\dot{\mathbf{v}}(t) = \frac{d\mathbf{v}(t)}{dt} = \left[\frac{dv_1(t)}{dt} \quad \frac{dv_2(t)}{dt} \quad \ldots \quad \frac{dv_n(t)}{dt} \right]^t \qquad (2.21)$$

$$\dot{\mathbf{v}}(t) = \begin{bmatrix} \lim_{\Delta t = 0} \frac{v_1(t+\Delta t)-v_1(t)}{\Delta t} \\ \vdots \\ \lim_{\Delta t = 0} \frac{v_n(t+\Delta t)-v_n(t)}{\Delta t} \end{bmatrix} \qquad (2.22)$$

As we have only one variable, the notation follows the total derivative $\mathrm{d}/\mathrm{d}t$ and often the literature uses an upper dot to denote the "time" derivative.

Graphically, the vector that results from the differentiation represents the vector tangent to the parametric plot of the function, as illustrated in Figure 2.3

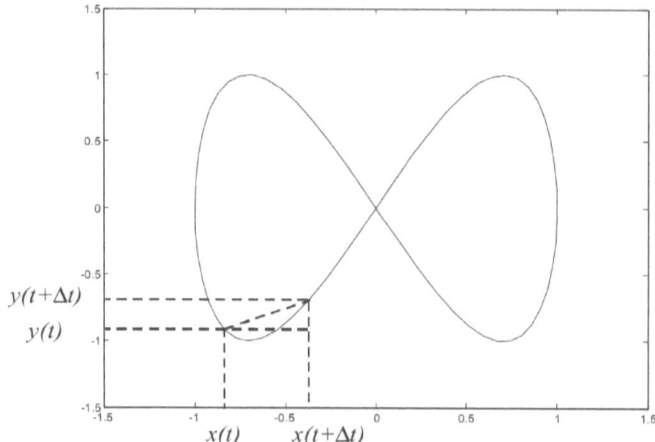

Figure 2.3: Line representing the derivative of a parametric function. In the limit $\Delta x \to 0$ the line will be tangent to the curve at the point $(x(t), y(t))$.

The Chain Rule is used in the situation where we have another function depending on t as

$$\mathbf{v}(t) = [\; v_1(\theta) \quad v_2(\theta) \quad \cdots \quad v_n(\theta) \;]^t$$
$$\theta = \theta(t) \qquad (2.23)$$

Since we have only one variable, the chain rule for differentiation is straightforward (the total derivative is a fraction) so we proceed exactly as we did in one function of one variable as

$$
\begin{aligned}
\frac{d\mathbf{v}(\theta(t))}{dt} &= \left[\begin{array}{cccc} \frac{dv_1}{dt}\frac{d\theta}{d\theta} & \frac{dv_2}{dt}\frac{d\theta}{d\theta} & \cdots & \frac{dv_n}{dt}\frac{d\theta}{d\theta} \end{array} \right]^t \\
&= \left[\begin{array}{cccc} \frac{dv_1}{d\theta}\frac{d\theta}{dt} & \frac{dv_2}{d\theta}\frac{d\theta}{dt} & \cdots & \frac{dv_n}{d\theta}\frac{d\theta}{dt} \end{array} \right]^t
\end{aligned}
\tag{2.24}
$$

and have the Chain Rule as ■

$$
\dot{\mathbf{v}}(\theta(t)) = \frac{d\mathbf{v}(\theta(t))}{dt} = \left[\begin{array}{cccc} \frac{dv_1}{d\theta}\dot{\theta} & \frac{dv_2}{d\theta}\dot{\theta} & \cdots & \frac{dv_n}{d\theta}\dot{\theta} \end{array} \right]^t
\tag{2.25}
$$

One function of several parametric functions

Consider the function $f(x_1(t), x_2(t))$. Its derivative with respect to time is (by definition)

$$
\dot{f}(x_1(t), x_2(t)) = \lim_{\Delta t \to 0} \frac{f(x_1(t+\Delta t), x_2(t+\Delta t)) - f(x_1(t), x_2(t))}{\Delta t}
\tag{2.26}
$$

For a small Δt we can write

$$
x_1(t+\Delta t) \approx x_1(t) + \dot{x}_1(t)\Delta t = x_1(t) + \Delta x_1(t)
\tag{2.27}
$$

Hence

$$
\frac{f(x_1(t)+\Delta x_1(t), x_2(t)+\Delta x_2(t)) - f(x_1(t), x_2(t))}{\Delta t} = \frac{\Delta f(x_1(t), x_2(t))}{\Delta t}
\tag{2.28}
$$

For small increments in x_1 and x_2 we can write (considering f to be continuous)

$$
f(x_1, x_2) = a\,x_1 + b\,x_2 + c
\tag{2.29}
$$

and with its increments we have

$$f(x_1, x_2) + \Delta f(x_1, x_2) = a(x_1 + \Delta x_1) + b(x_2 + \Delta x_2) + c$$
$$= ax_1 + a\Delta x_1 + bx_2 + b\Delta x_2 + c$$
$$= f(x_1, x_2) + a\Delta x_1 + b\Delta x_2$$

$$(2.30)$$

hence

$$\Delta f(x_1, x_2) = a\Delta x_1 + b\Delta x_2 \qquad (2.31)$$

For continuous functions and points very close to x_1 and x_2, we have (slopes in each direction)

$$a = \frac{\partial f(x_1, x_2)}{\partial x_1}, b = \frac{\partial f(x_1, x_2)}{\partial x_2} \qquad (2.32)$$

Hence

$$\Delta f(x_1(t), x_2(t)) = \frac{\partial f(x_1(t), x_2(t))}{\partial x_1(t)} \Delta x_1(t) + \frac{\partial f(x_1(t), x_2(t))}{\partial x_2(t)} \Delta x_2(t)$$

$$(2.33)$$

Using 2.27 we write

$$\dot{x}_1(t)\Delta t = \Delta x_1(t) \qquad (2.34)$$

Now, substituting 2.33 in 2.28 we have

$$\frac{\Delta f(x_1(t), x_2(t))}{\Delta t} = \frac{\frac{\partial f(x_1(t), x_2(t))}{\partial x_1(t)} \Delta x_1(t) + \frac{\partial f(x_1(t), x_2(t))}{\partial x_2(t)} \Delta x_2(t)}{\Delta t}$$

$$(2.35)$$

and substituting 2.34 in 2.35 we have

$$\dot{f}(x_1(t), x_2(t)) = \lim_{\Delta t \to 0} \frac{\Delta f(x_1(t), x_2(t))}{\Delta t}$$
$$= \lim_{\Delta t \to 0} \frac{\frac{\partial f(x_1(t), x_2(t))}{\partial x_1(t)} \dot{x}_1(t)\Delta t + \frac{\partial f(x_1(t), x_2(t))}{\partial x_2(t)} \dot{x}_2(t)\Delta t}{\Delta t} \qquad (2.36)$$

which finally leads to ■

$$\dot{f}(x_1(t), x_2(t)) = \frac{\partial f(x_1, x_2)}{\partial x_1}\dot{x}_1(t) + \frac{\partial f(x_1, x_2)}{\partial x_2}\dot{x}_2(t) \quad (2.37)$$

2.2.4 Several functions of several variables

Now, consider the vector representation of several functions depending on several variables. Formally

$$\begin{bmatrix} v_1 \\ \vdots \\ v_n \end{bmatrix} = \begin{bmatrix} f_1(x_1, x_2, \ldots, x_m) \\ \vdots \\ f_n(x_1, x_2, \ldots, x_m) \end{bmatrix} \quad (2.38)$$

We can write that in a compact form as

$$\mathbf{v} = \mathbf{f}(\mathbf{x}) \quad (2.39)$$

where $\mathbf{v} = \begin{bmatrix} v_1 & \cdots & v_n \end{bmatrix}^t$ and $\mathbf{x} = \begin{bmatrix} x_1 & \cdots & x_m \end{bmatrix}^t$.

Using the results from the previous section for each function we write the differentiation in this case as

$$\begin{aligned} dv_1 &= \frac{\partial f_1}{\partial x_1}dx_1 + \frac{\partial f_1}{\partial x_2}dx_2 + \cdots + \frac{\partial f_1}{\partial x_m}dx_m \\ dv_2 &= \frac{\partial f_2}{\partial x_1}dx_1 + \frac{\partial f_2}{\partial x_2}dx_2 + \cdots + \frac{\partial f_2}{\partial x_m}dx_m \\ &\qquad\qquad\qquad \vdots \\ dv_n &= \frac{\partial f_n}{\partial x_1}dx_1 + \frac{\partial f_n}{\partial x_2}dx_2 + \cdots + \frac{\partial f_n}{\partial x_m}dx_m \end{aligned} \quad (2.40)$$

And, again, we can write it in a more compact form as

$$d\mathbf{v} = \mathbf{J}(\mathbf{v})d\mathbf{x} \quad (2.41)$$

with

$$\mathbf{J}(\mathbf{x}) = \begin{bmatrix} \frac{\partial f_1}{\partial x_1} & \frac{\partial f_1}{\partial x_2} & \cdots & \frac{\partial f_1}{\partial x_m} \\ \frac{\partial f_2}{\partial x_1} & \frac{\partial f_2}{\partial x_2} & & \frac{\partial f_2}{\partial x_m} \\ \vdots & & \ddots & \vdots \\ \frac{\partial f_n}{\partial x_1} & \frac{\partial f_n}{\partial x_2} & \cdots & \frac{\partial f_n}{\partial x_m} \end{bmatrix} \quad (2.42)$$

and $d\mathbf{x}$ and $d\mathbf{v}$ the differentials for each element in $\mathbf{x}$ and $\mathbf{v}$ ■.

Several functions of several parametric functions

Now, consider the case of

$$\mathbf{v} = \left[\begin{array}{c} f_1(x_1(t), x_2(t)) \\ f_2(x_1(t), x_2(t)) \end{array} \right] \tag{2.43}$$

Using the definition for the derivative, we have

$$\dot{\mathbf{v}} = \frac{\mathrm{d}\mathbf{v}}{\mathrm{d}t} = \left[\begin{array}{c} \frac{\mathrm{d}f_1(x_1(t),x_2(t))}{\mathrm{d}t} \\ \frac{\mathrm{d}f_2(x_1(t),x_2(t))}{\mathrm{d}t} \end{array} \right] \tag{2.44}$$

Using previous results for functions of several parametric functions, we have

$$\left[\begin{array}{c} \frac{\mathrm{d}f_1(x_1(t),x_2(t))}{\mathrm{d}t} \\ \frac{\mathrm{d}f_2(x_1(t),x_2(t))}{\mathrm{d}t} \end{array} \right] = \left[\begin{array}{c} \frac{\partial f_1(x_1,x_2)}{\partial x_1}\dot{x}_1(t) + \frac{\partial f_1(x_1,x_2)}{\partial x_2}\dot{x}_2(t) \\ \frac{\partial f_2(x_1,x_2)}{\partial x_1}\dot{x}_1(t) + \frac{\partial f_2(x_1,x_2)}{\partial x_2}\dot{x}_2(t) \end{array} \right] \tag{2.45}$$

That can be written in matrix form as

$$\left[\begin{array}{c} \frac{\partial f_1(x_1,x_2)}{\partial x_1}\dot{x}_1(t) + \frac{\partial f_1(x_1,x_2)}{\partial x_2}\dot{x}_2(t) \\ \frac{\partial f_2(x_1,x_2)}{\partial x_1}\dot{x}_1(t) + \frac{\partial f_2(x_1,x_2)}{\partial x_2}\dot{x}_2(t) \end{array} \right] = $$
$$\left[\begin{array}{cc} \frac{\partial f_1(x_1,x_2)}{\partial x_1} & \frac{\partial f_1(x_1,x_2)}{\partial x_2} \\ \frac{\partial f_2(x_1,x_2)}{\partial x_1} & \frac{\partial f_2(x_1,x_2)}{\partial x_2} \end{array} \right] \left[\begin{array}{c} \dot{x}_1(t) \\ \dot{x}_2(t) \end{array} \right] \tag{2.46}$$

That leads to the compact form ■

$$\dot{\mathbf{v}} = \mathbf{J}(\mathbf{x}(t))\dot{\mathbf{x}}(t) \tag{2.47}$$

2.2.5 Total derivative versus partial derivative

To summarise the importance of the notation for total and partial derivative, let us analyse the difference and relationship between the two. In the previous sections we showed that if we have a function of several parametric functions, we have the relation

$$\frac{\mathrm{d}f}{\mathrm{d}t} = \frac{\partial f}{\partial x_1}\frac{\mathrm{d}x_1}{\mathrm{d}t} + \frac{\partial f}{\partial x_2}\frac{\mathrm{d}x_2}{\mathrm{d}t} + \frac{\partial f}{\partial x_3}\frac{\mathrm{d}x_3}{\mathrm{d}t} \tag{2.48}$$

This means that we are measuring the total variation for the function f letting all variables x_1, x_2 and x_3 vary when we change t. Hence, the variations $\mathrm{d}x_i$ are numbers (since each function $x_i(t)$ depends only on t). Moreover, we are assuming that f has no direct dependency on t (it depends exclusively though on x_i). With that in mind, we can divide both sides by $\mathrm{d}t$ and get

$$\mathrm{d}f = \frac{\partial f}{\partial x_1}\mathrm{d}x_1 + \frac{\partial f}{\partial x_2}\mathrm{d}x_2 + \frac{\partial f}{\partial x_3}\mathrm{d}x_3 \tag{2.49}$$

Now, we divide for the variation of one of the variables (say x_1) and get

$$\frac{\mathrm{d}f}{\mathrm{d}x_1} = \frac{\partial f}{\partial x_1} + \frac{\partial f}{\partial x_2}\frac{\mathrm{d}x_2}{\mathrm{d}x_1} + \frac{\partial f}{\partial x_3}\frac{\mathrm{d}x_3}{\mathrm{d}x_1} \tag{2.50}$$

Now, without loss of generality let us call x_1 as t and generalise for n variables. We get

$$\begin{aligned}
\frac{\mathrm{d}f}{\mathrm{d}t} &= \frac{\partial f}{\partial t} + \frac{\partial f}{\partial x_1}\frac{\mathrm{d}x_1}{\mathrm{d}t} + \frac{\partial f}{\partial x_3}\frac{\mathrm{d}x_2}{\mathrm{d}t} + \dots \\
\frac{\mathrm{d}f}{\mathrm{d}t} &= \frac{\partial f}{\partial t} + \sum_{i=1}^{n} \frac{\partial f}{\partial x_i}\frac{\mathrm{d}x_i}{\mathrm{d}t}
\end{aligned} \tag{2.51}$$

which is clarifying evidence that $\frac{\partial f}{\partial x}$ is completely different from $\frac{\mathrm{d}f}{\mathrm{d}x}$ ■.

2.3 Integration

In this section we are going to visit the topic of integration. We will start with the definition (proving it) and investigate a few cases of integration like integration by parts and arc-length.

2.3.1   Definition and anti-derivative

We can define the integral of a function f from a point a to b as the area under the function (allowing the area being negative if the function is negative). This area can be computed as the limit of an approximation given by rectangles positioned at x_i (from a to b) with base length Δx and height equal to $f(x_i)$. Figure 2.4 illustrate this approximation.

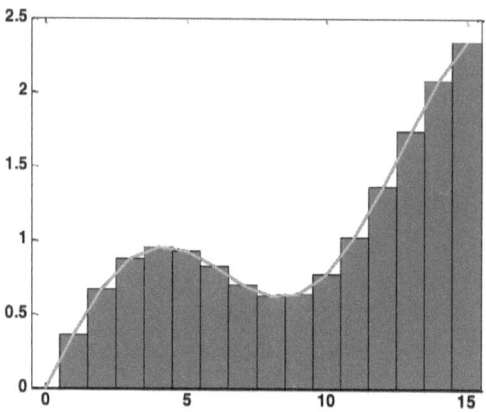

Figure 2.4: Representation of the finite approximation for the area under a function f

In the limit $(\Delta x \to 0)$ the area will be exact and could be calculated by

$$\int_a^b f(x)\mathrm{d}x = \lim_{N\to\infty} \sum_{i=0}^{N} f\left(a + \frac{b-a}{N}i\right)\frac{b-a}{N} \qquad (2.52)$$

As in the derivative of a function, we do not always have to compute the limit by the definition. We could try to figure out a systematic way to compute the value of the integral. Let us compute the integral from 0 to the limits and call that a function of the limit as

$$\int_0^a f(x)dx = \lim_{N\to\infty} \sum_{i=0}^{N} f\left(\frac{a}{N}i\right)\frac{a}{N} = F(a) \tag{2.53}$$

and

$$\int_0^a f(x)dx = \lim_{N\to\infty} \sum_{i=0}^{N} f\left(\frac{a}{N}i\right)\frac{a}{N} = F(a) \tag{2.54}$$

As $N \to \infty$ the area up to a in the first integral will coincide with the area of the second and we can compute the integral as

$$\int_a^b f(x)dx = F(b) - F(a) \tag{2.55}$$

Now, we have to figure out a way to compute $F(x)$. To do that we analyse the following approximations

$$F(x) \approx \sum_{i=0}^{N} f(i\Delta x)\Delta x \tag{2.56}$$

and

$$F(x + \Delta x) \approx \sum_{i=0}^{N+1} f(i\Delta x)\Delta x \tag{2.57}$$

Now, we subtract the two and get

$$F(x + \Delta x) - F(x) \approx \sum_{i=0}^{N+1} f(i\Delta x)\Delta x - \sum_{i=0}^{N} f(i\Delta x)\Delta x \tag{2.58}$$

As the first terms of the summations cancel each other out and $f((N+1)\Delta x)\Delta x \approx f(x)\Delta x$, we are left with

$$f(x) \approx \frac{F(x + \Delta x) - F(x)}{\Delta x} \tag{2.59}$$

Hence, in the limit we have

$$f(x) = \frac{\mathrm{d}F(x)}{\mathrm{d}x} \tag{2.60}$$

This is sometimes called the Fundamental Theorem Of Calculus. And amazingly relates the area under a function to its "anti-derivative" ■.

2.3.2 Integration by parts

Using the result from section 2.2.1 we can derive an alternative way to integrate the product of two functions. This is called the Product Rule and we can derive that rule by starting with

$$\frac{\mathrm{d}(u(t)v(t))}{\mathrm{d}t} = \frac{v(t)\mathrm{d}u(t) + u(t)\mathrm{d}v(t)}{\mathrm{d}t}$$
$$\mathrm{d}(u(t)v(t)) = v(t)\mathrm{d}u(t) + u(t)\mathrm{d}v(t) \tag{2.61}$$

Using a more compact form (since the dependency on t is common to all terms), we have

$$\mathrm{d}(uv) = v\mathrm{d}u + u\mathrm{d}v \tag{2.62}$$

Now, integrating both sides we have

$$\int_0^t \mathrm{d}(uv) = \int_0^t v\mathrm{d}u + \int_0^t u\mathrm{d}v \tag{2.63}$$

That results in

$$uv\big|_0^t = \int_0^t v\mathrm{d}u + \int_0^t u\mathrm{d}v \tag{2.64}$$

which leads us to the useful Product Rule as ■

$$\int_0^t u\mathrm{d}v = uv\big|_0^t - \int_0^t v\mathrm{d}u \tag{2.65}$$

2.3.3 Arc-length

Another common use for integrals is to calculate the length of a curve. In particular, it is possible to calculate the length of a parametric curve of the type

$$\mathbf{f}(t) = \begin{bmatrix} x(t) \\ y(t) \end{bmatrix} \tag{2.66}$$

with t ranging from t_0 to t_1. We proceed with an approximation, similar to the way it is done with normal integrals. That would approximate the curve by line segments, as indicated in Figure 2.5

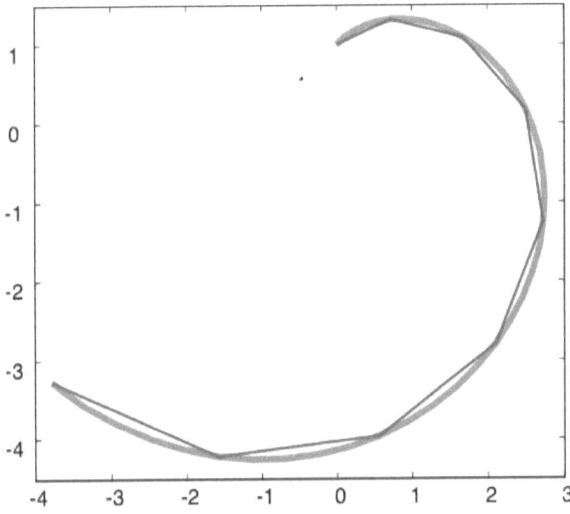

Figure 2.5: Representation of the finite approximation for the arc length of a parametric function f

The approximation can now be computed as

$$C = \lim_{N \to \infty} \sum_{i=0}^{N} \sqrt{(\Delta x_i)^2 + (\Delta y_i)^2} \tag{2.67}$$

where

$$\begin{aligned} \Delta x_i &= \tfrac{x(t_0+(i+1)\Delta t)-x(t_0+i\Delta t)}{N} \\ \Delta y_i &= \tfrac{y(t_0+(i+1)\Delta t)-y(t_0+i\Delta t)}{N} \\ \Delta t &= \tfrac{t_1-t_0}{N} \end{aligned} \tag{2.68}$$

If we divide and multiply by Δt inside the summation, we have

$$C = \lim_{N\to\infty} \sum_{i=0}^{N} \sqrt{\left(\frac{\Delta x_i}{\Delta t}\right)^2 + \left(\frac{\Delta y_i}{\Delta t}\right)^2}\,\Delta t \tag{2.69}$$

which in the limit becomes ■

$$C = \int_{t_0}^{t_1} \sqrt{\dot{x}^2(t) + \dot{y}^2(t)}\,\mathrm{d}t \tag{2.70}$$

2.4   Differentiation under integral

We are interested in evaluating the derivative of the following expression

$$H(x) = \int_{a(x)}^{b(x)} h(x,y)\mathrm{d}y \tag{2.71}$$

where the functions $h(x,y)$, $a(x)$ and $b(x)$ are known. As $H(x)$ depends only on x, we would like to find an expression to $\frac{\mathrm{d}H(x)}{\mathrm{d}x}$. Let us start by taking the differential approach and write the increments in $H(x)$ with a small increment in x. If we write all functions that depend on x as their increment, we get

$$H(x) + \Delta H(x) = \int_{a(x)+\Delta a(x)}^{b(x)+\Delta b(x)} \left(h(x,y) + \Delta h(x,y)\right)\mathrm{d}y \tag{2.72}$$

Now, we identify the new integral as a "displaced" version of the original one (and slightly changed in shape) as Figure 2.6 shows.

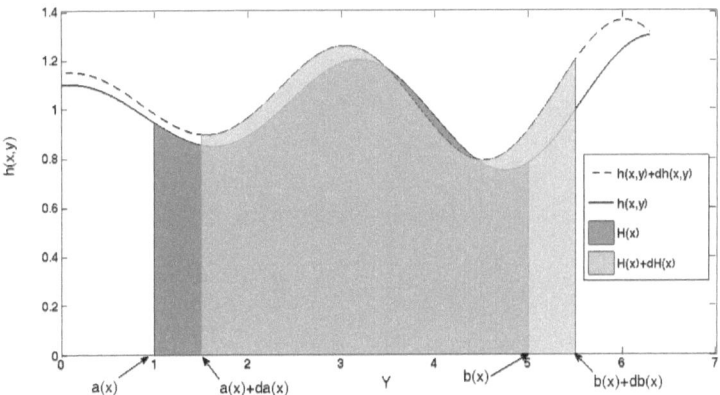

Figure 2.6: Representation of the area for the original integral (eq. 2.71) and the incremented one (eq. 2.72).

In the differential approach, we must rewrite the right-hand side of (2.72) so as to force the appearance of a term equal to $H(x)$ (eq. 2.71). By doing that we are left with an expression for $\Delta H(x)$ that can be used to write the derivative. Since our integral in $H(x)$ goes from $a(x)$ to $b(x)$, we will add and subtract the area from $a(x)$ to $a(x) + \Delta a(x)$ from equation 2.72. We will be left with

$$H(x) + \Delta H(x) = \int_{a(x)+\Delta a(x)}^{b(x)+\Delta b(x)} \left(h(x,y) + \Delta h(x,y)\right) dy +$$

$$\int_{a(x)}^{a(x)+\Delta a(x)} \left(h(x,y) + \Delta h(x,y)\right) dy - \int_{a(x)}^{a(x)+\Delta a(x)} \left(h(x,y) + \Delta h(x,y)\right) dy$$

$$= \int_{a(x)}^{b(x)+\Delta b(x)} \left(h(x,y) + \Delta h(x,y)\right) dy - \int_{a(x)}^{a(x)+\Delta a(x)} \left(h(x,y) + \Delta h(x,y)\right) dy$$

$$(2.73)$$

That can also be written by (separating the last piece of area from $b(x)$ to $b(x) + \Delta b(x)$)

$$H(x) + \Delta H(x) = \int_{a(x)}^{b(x)} \left(h(x,y) + \Delta h(x,y)\right) dy +$$

$$\int_{b(x)}^{b(x)+\Delta b(x)} \left(h(x,y) + \Delta h(x,y)\right) dy - \int_{a(x)}^{a(x)+\Delta a(x)} \left(h(x,y) + \Delta h(x,y)\right) dy$$

$$= \int_{a(x)}^{b(x)} h(x,y)dy + \int_{a(x)}^{b(x)} \Delta h(x,y)dy + \int_{b(x)}^{b(x)+\Delta b(x)} h(x,y)dy +$$

$$\int_{b(x)}^{b(x)+\Delta b(x)} \Delta h(x,y)dy - \int_{a(x)}^{a(x)+\Delta a(x)} h(x,y)dy - \int_{a(x)}^{a(x)+\Delta a(x)} \Delta h(x,y)dy$$

$$(2.74)$$

Now, we eliminate $H(x)$ from both sides and are left with

$$\Delta H(x) = \int_{a(x)}^{b(x)} \Delta h(x,y) dy + \int_{b(x)}^{b(x)+\Delta b(x)} h(x,y) dy + \int_{b(x)}^{b(x)+\Delta b(x)} \Delta h(x,y) dy -$$

$$\int_{a(x)}^{a(x)+\Delta a(x)} h(x,y) dy - \int_{a(x)}^{a(x)+\Delta a(x)} \Delta h(x,y) dy$$

$$(2.75)$$

Now, since $H(x)$ is a primitive of $h(x,y)$ in y we can integrate the second and fourth terms on the right-hand side of (2.75). We can also define a primitive $\delta H(x,y)$ to $\Delta h(x,y)$ and integrate the third and fifth terms, leading us to

$$\Delta H(x) = \int_{a(x)}^{b(x)} \Delta h(x,y) dy + (H(x,b(x)+\Delta b(x)) - H(x,b(x)))$$

$$+ (\delta H(b(x)+\Delta b(x), b(x)) - \delta H(b(x), b(x)))$$
$$- (H(x,a(x)+\Delta a(x)) - H(x,a(x)))$$
$$- (\delta H(x,a(x)+\Delta a(x)) - \delta H(x,a(x)))$$

$$(2.76)$$

As $\Delta h(x,y)$ will tend to zero, so will its primitive $\delta H(x,y)$. Now, we are left with

$$\Delta H(x) = \int_{a(x)}^{b(x)} \Delta h(x,y) dy + (H(x,b(x)+\Delta b(x)) - H(x,b(x))) -$$

$$(H(x,a(x)+\Delta a(x)) - H(x,a(x)))$$

$$(2.77)$$

Now, we are ready to divide both sides by Δx and take the limit $\Delta x \to 0$. This will write the left-hand side and the first term of the right-hand side of (2.77) as differentials. On the other hand, we still can't write the differences of the right-hand side as differentials because they depend on $b(x)$ and $a(x)$ respectively

$$\frac{\Delta H(x)}{\Delta x} = \int_{a(x)}^{b(x)} \frac{\Delta h(x,y)}{\Delta x} dy +$$

$$\frac{H(x,b(x)+\Delta b(x)) - H(x,b(x))}{\Delta x} - \frac{H(x,a(x)+\Delta a(x)) - H(x,a(x))}{\Delta x}$$

$$(2.78)$$

Nevertheless, we can multiply the second and third term on the right-hand side of (2.78) by $\Delta a(x)/\Delta a(x)$ and $\Delta b(x)/\Delta b(x)$ respectively, leading us to

$$\frac{\Delta H(x)}{\Delta x} = \int_{a(x)}^{b(x)} \frac{\Delta h(x,y)}{\Delta x} dy +$$

$$\frac{H(x,b(x)+\Delta b(x)) - H(x,b(x))}{\Delta b(x)} \frac{\Delta b(x)}{\Delta x} - \qquad (2.79)$$

$$\frac{H(x,a(x)+\Delta a(x)) - H(x,a(x))}{\Delta a(x)} \frac{\Delta a(x)}{\Delta x}$$

Now, taking the limit, we have

$$\frac{dH(x)}{dx} = \int_{a(x)}^{b(x)} \frac{dh(x,y)}{dx} dy + h(x,b(x))\frac{db(x)}{dx} - h(x,a(x))\frac{da(x)}{dx}$$

$$(2.80)$$

which is the differentiation under integral due to Leibniz ∎.

2.5   Numeric methods

2.5.1   Simpson's Rule

A very useful numerical tool that can be used in many applications is the numerical integration. In this section we will derive the Simpson's formula for solving the following integral

$$I = \int_a^b f(x)\mathrm{d}x \tag{2.81}$$

Where, instead of $f(x)$, we have a set of equally spaced points $((x_0, y_0), (x_1, y_1), \cdots, (x_n, y_n))$ where x_i has values from a to b as $a, a+h, a+2\,h, \cdots, b$ and y_i has values $f(a)$, $f(a+h)$, $f(a+2h)$, $\cdots$, $f(b)$.

The idea behind the Simpson's Rule is to divide the set of (x_i, y_i) in groups of n pairs and for each set, interpolate a polynomial. Then we write directly the integral of the polynomial. Here we will derive the second rule (2nd degree polynomial) with the interpolation

$$ax_i^2 + bx_i + c = y_i \tag{2.82}$$

For each polynomial, we normalise the x_i to be $0, h, 2\,h$ since that won't change the value of the integral as long as we the corresponding y_i. So, we have the following interpolating equations

$$\begin{aligned}
a \cdot 0 + b \cdot 0 + c &= y_0 \\
ah^2 + bh + c &= y_1 \\
a(2h)^2 + b(2h) + c &= y_2
\end{aligned} \tag{2.83}$$

Solving for a, b and c we get

$$\begin{aligned}
a &= \frac{y_0 - 2\,y_1 + y_2}{2\,h^2} \\
b &= -\frac{3\,y_0 - 4\,y_1 + y_2}{2\,h} \\
c &= y_0
\end{aligned} \tag{2.84}$$

Now, the integral of the set is

$$I = \int_0^{2h} \left(ax^2 + bx + c\right)dx \tag{2.85}$$

$$= \left. \frac{1}{3}ax^3 + \frac{1}{2}bx^2 + cx \right|_0^{2h}$$

That results in

$$I = \frac{8}{3}ah^3 + \frac{9}{2}bh^2 + 2ch$$

$$I = \frac{8}{3}\frac{y_0 - 2\,y_1 + y_2}{2\,h^2}h^3 - \frac{9}{2}\frac{3\,y_0 - 4\,y_1 + y_2}{2\,h}h^2 + 2y_0h \tag{2.86}$$

and leads to

$$I = \frac{h}{3}\left(y_0 + 4y_1 + y_2\right) \tag{2.87}$$

That is the interpolating integral for the first set of three points. Now, we add all the sets and get (notice that the first of each following set is also the last of the preceding one, hence the "2" appearing in the formula) ■.

$$I = \frac{h}{3}\left(y_0 + 4y_1 + 2y_2 + 4y_3 + 2y_4 + \ldots + 2y_{n-2} + 4y_{n-1} + y_n\right) \tag{2.88}$$

2.5.2   Newton's Method

In this section we will derive the Newton's method for finding roots. Formally, we want the values of x that satisfy

$$0 = f(x) \tag{2.89}$$

The idea behind Newton's method is to approximate the function $f(x)$ by an affine function of the form

$$0 \approx a\,x + b \tag{2.90}$$

and find the value x. In this approximation, the angular coefficient a is given by the derivative of $f(x)$ and b can be found by setting the affine function to pass to an initial point x_0 as Figure 2.7 illustrates.

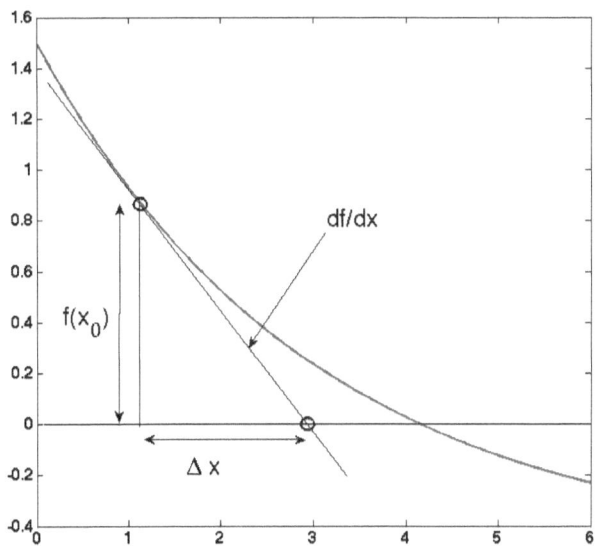

Figure 2.7: Graphical representation for one iteration of Newton's Method

Hence, we have

$$a = \left. \frac{\mathrm{d}f(x)}{\mathrm{d}x} \right|_{x_0} = f'(x_0) \qquad (2.91)$$

and b can be found by

$$\begin{aligned} f(x_0) &= f'(x_0)x_0 + b \\ b &= f(x_0) - f'(x_0)x \end{aligned} \qquad (2.92)$$

Now, solving

$$0 = ax + b \qquad (2.93)$$

we have

$$x = -\frac{b}{a}$$

$$= -\frac{f(x_0) - f'(x_0)x_0}{f'(x_0)} \tag{2.94}$$

$$= -\frac{f(x_0)}{f'(x_0)} + x_0$$

Since we want an iterative process, we need to find a Δx to set the next point to approximate the solution. Hence

$$\Delta x = x - x_0$$

$$= -\frac{f(x_0)}{f'(x_0)} \tag{2.95}$$

and finally we have Newton's Method as the following iterative equation ■

$$x[n + 1] = x[n] - \frac{f(x[n])}{f'(x[n])} \tag{2.96}$$

2.5.3   Poison's Equation

The Poison's Equation is a partial differential equation of the form

$$\nabla^2 f(x, y) = g(x, y) \tag{2.97}$$

where $f(x, y)$ is the function we wish to find and $g(x, y)$ is a known function. The ∇^2 operator is the Laplacian operator. The Laplacian of a two-variable function is given by

$$\nabla^2 = \frac{\partial^2}{\partial x} + \frac{\partial^2}{\partial y} \tag{2.98}$$

In the numeric solution of equation 2.97, we are interested in finding the values of $f(x, y)$ for a given set of discretised points (x_i, y_i). In order to do that, we must approximate the Laplacian of a function at a point P using its neighbours in the horizontal and vertical directions. Figure 1 shows the structure of the points used in this approximation

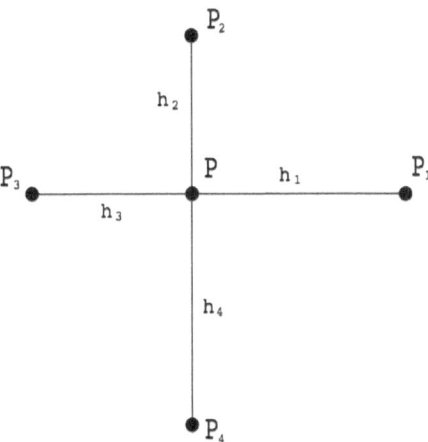

Figure 2.8: Discretisation used in the numeric solution of the Poison Equation

Figure 2.8 shows a point $P = (x_i, y_i)$ and its four neighbours $(P_1, P_2, P_3$ and $P_4)$. The distances from P to its neighbours are h_1, h_2, h_3 and h_4. We will call the value of $f(x, y)$ at P as v and the value at its neighbours as v_1, v_2, v_3 and v_4. The goal is to find an expression for v depending on v_1 to v_4, h_1 to h_4 and $g(x_i, y_i)$. To do that we must find an approximation to the second derivative at the point P. Let

$$f_2(x, y) = \frac{\partial f(x, y)}{\partial x} \qquad (2.99)$$

Hence, we have

$$\frac{\partial^2 f(x, y)}{\partial x^2} = \frac{\partial f_2(x, y)}{\partial x} \qquad (2.100)$$

We can approximate the derivative in the x-direction as

$$\frac{\partial f(x, y)}{\partial x} \approx \frac{f(x + \Delta x, y) - f(x, y)}{\Delta x} \qquad (2.101)$$

If we use that to approximate $f_2(x, y)$, we have

$$\frac{\partial f_2(x, y)}{\partial x} \approx \frac{f_2(x, y) - f_2(x - h_3, y)}{h_3} \tag{2.102}$$

which is an approximation to $\nabla^2 f(x, y)$ at P. Again, we can approximate $f_2(x, y)$ and $f_2(x - h_3, y)$ (at the points P and P_3 respectively) as

$$f_2(x, y) = \frac{\partial f(x, y)}{\partial x} \approx \frac{f(x + h_1, y) - f(x, y)}{h_1} \tag{2.103}$$

and

$$f_2(x - h_3, y) = \frac{\partial f(x - h_3, y)}{\partial x} \approx \frac{f(x, y) - f(x - h_3, y)}{h_3} \tag{2.104}$$

Substituting equation 2.104 in equation 2.102 we have

$$\frac{\partial^2 f(x, y)}{\partial x^2} = \frac{\partial f_2(x, y)}{\partial x} \approx \frac{\frac{f(x+h_1,y)-f(x,y)}{h_1} - \frac{f(x,y)-f(x-h_3,y)}{h_3}}{h_3} \tag{2.105}$$

Hence, we have

$$\frac{\partial^2 f(x, y)}{\partial x^2} \approx \frac{h_3 f(x + h_1, y) - (h_3 + h_1) f(x, y) + h_1 f(x - h_3, y)}{h_1 h_3^2} \tag{2.106}$$

We can do the same procedure to the approximation in y and obtain

$$\frac{\partial^2 f(x, y)}{\partial y^2} \approx \frac{h_4 f(x, y + h_2) - (h_4 + h_2) f(x, y) + h_2 f(x, y - h_4)}{h_2 h_4^2} \tag{2.107}$$

Now the Poison's Equation is then approximated as

$$\frac{h_3 v_1 - (h_3 + h_1)v + h_1 v_3}{h_1 h_3^2} + \frac{h_4 v_2 - (h_4 + h_2)v + h_2 v_4}{h_2 h_4^2} \approx g(x, y) \tag{2.108}$$

where

$$v = f(x, y)$$
$$v_1 = f(x + h_1, y)$$
$$v_2 = f(x, y + h_2) \tag{2.109}$$
$$v_3 = f(x - h_3, y)$$
$$v_4 = f(x, y - h_4)$$

Solving for v, we have the linear approximation for the Poison's Equation at the point P written as

$$v \approx \frac{h_2 h_3 h_4^2 v_1 + h_2 h_1 h_4^2 v_3 + h_1 h_4 h_3^2 v_2 + h_1 h_2 h_3^2 v_4 - h_1 h_2 h_3^2 h_4^2 g(x, y)}{h_2 h_4^2 (h_3 + h_1) + h_1 h_3^2 (h_2 + h_4)}$$

$$\tag{2.110}$$

2.6   Vector Differentiation

In this section we will derive some basic but important results from vector differentiation. This kind of differential manipulation can be very useful in proving and deriving important results in almost all fields of engineering. Normally, the student is not familiar with some of the concepts of vector calculus. For instance the derivative $\frac{\partial \|x\|^2}{\partial x}$ when $\mathbf{x}$ is a vector. However, this is a very precise quantity that means how much the norm of the vector changes when the vector itself changes. As we will see, that quantity is a vector, but more complicated cases exist and in this book we do not even scratch the surface of the subject. Instead, we will try to give some very basic results that allow the readers to perform their own derivation.

2.6.1   Definitions

Before any derivation, it is necessary to establish some definitions. In vector differentiation, we must define the direction we write the vector. That is important because we need quantities like $\frac{\partial x^t s}{\partial x}$ and $\frac{\partial s^t x}{\partial x}$ to be precisely defined. Some literature calls that definition a "numerator layout" or "denominator layout", depending on how we define the vectors as columns or rows. In this section we will use the numerator layout and we have the following definitions for matrix and vectors

$$\mathbf{A} = \begin{bmatrix} a_{1,1} & a_{1,2} & \cdots & a_{1,n} \\ a_{2,1} & a_{2,2} & & a_{2,n} \\ \vdots & & \ddots & \\ a_{m,1} & a_{m,2} & & a_{m,n} \end{bmatrix} \qquad (2.111)$$

$$\mathbf{x} = \begin{bmatrix} x_1 \\ x_2 \\ \vdots \\ x_n \end{bmatrix} \qquad (2.112)$$

When possible, matrices will be written in uppercase bold and vectors as lowercase bold. We will define two cases for differentiation. We can have scalar differentiation when we have a function $f : \Re^n \to \Re$. In that case we use the following definition

$$\frac{\partial f(\mathbf{x})}{\partial \mathbf{x}} = \begin{bmatrix} \frac{\partial f(\mathbf{x})}{\partial x_1} & \frac{\partial f(\mathbf{x})}{\partial x_2} & \cdots & \frac{\partial f(\mathbf{x})}{\partial x_n} \end{bmatrix} \qquad (2.113)$$

where $f(.)$ takes a vector and returns a scalar. Notice that we are using the layout definition that differentiates as a row vector (but $\mathbf{x}$ is a column vector). The second definition is the case for $\mathbf{f} : \Re^n \to \Re^m$. In that case we have

$$\frac{\partial \mathbf{f}(\mathbf{x})}{\partial \mathbf{x}} = \begin{bmatrix} \frac{\partial f_1(\mathbf{x})}{\partial x_1} & \frac{\partial f_1(\mathbf{x})}{\partial x_2} & \cdots & \frac{\partial f_1(\mathbf{x})}{\partial x_n} \\ \frac{\partial f_2(\mathbf{x})}{\partial x_1} & \frac{\partial f_2(\mathbf{x})}{\partial x_2} & & \frac{\partial f_2(\mathbf{x})}{\partial x_n} \\ \vdots & & \ddots & \\ \frac{\partial f_m(\mathbf{x})}{\partial x_1} & \frac{\partial f_m(\mathbf{x})}{\partial x_2} & & \frac{\partial f_m(\mathbf{x})}{\partial x_n} \end{bmatrix} \qquad (2.114)$$

and it is important to notice the result is a matrix. It is also important to notice the direction of the elements in the matrix. Each row is the differentiation of one of the functions for $\mathbf{f}$.

In the case when we have a vector differentiated by a scalar, we proceed as with a set of scalar functions and differentiate each element of the vector as

$$
\frac{\partial \mathbf{f}(\mathbf{x})}{\partial x_i} =
\begin{bmatrix}
\frac{\partial \mathbf{f}(\mathbf{x})}{\partial x_1} \\
\frac{\partial \mathbf{f}(\mathbf{x})}{\partial x_2} \\
\vdots \\
\frac{\partial \mathbf{f}(\mathbf{x})}{\partial x_n}
\end{bmatrix}
\tag{2.115}
$$

2.6.2 Inner product

Let us see the derivative of a scalar function of the form

$$
f(\mathbf{x}) = \mathbf{x}^t \mathbf{s}
\tag{2.116}
$$

As we know by the layout of the result by the definitions above, we need to compute each component of the resulting vector. We do that differentiating f for each component of the vector (as the definition states)

$$
\frac{\partial f(\mathbf{x})}{\partial x_k} = s_k
\tag{2.117}
$$

If we write the resulting vector, we get ■

$$
\frac{\partial f(\mathbf{x})}{\partial \mathbf{x}} = \mathbf{s}^t
\tag{2.118}
$$

2.6.3 Product rule for vector calculus

Now, we will derive some more effective results. We are now interested in differentiating scalar functions of the form

$$
f(\mathbf{x}) = \mathbf{h}^t(\mathbf{x}) \mathbf{s}(\mathbf{x})
\tag{2.119}
$$

To do that, we first write f as function of each individual element for the vector function as

$$f(\mathbf{x}) = \sum_i h_i(\mathbf{x}) s_i(\mathbf{x}) \tag{2.120}$$

Differentiating with respect to each element of $\mathbf{x}$, we have

$$
\begin{aligned}
\frac{\partial f(\mathbf{x})}{\partial x_j} &= \sum_i \frac{\partial \left(h_i(\mathbf{x}) s_i(\mathbf{x}) \right)}{\partial x_j} \\
&= \sum_i \frac{\partial h_i(\mathbf{x})}{\partial x_j} s_i(\mathbf{x}) + h_i(\mathbf{x}) \frac{\partial s_i(\mathbf{x})}{\partial x_j}
\end{aligned}
\tag{2.121}
$$

We can write a more compact form for 2.121 if we write the summation as

$$
\frac{\partial f(\mathbf{x})}{\partial x_j} = \begin{bmatrix} s_1(\mathbf{x}) & s_2(\mathbf{x}) & \cdots & s_n(\mathbf{x}) \end{bmatrix}
\begin{bmatrix} \frac{\partial h_1(\mathbf{x})}{\partial x_j} \\ \frac{\partial h_2(\mathbf{x})}{\partial x_j} \\ \vdots \\ \frac{\partial h_n(\mathbf{x})}{\partial x_j} \end{bmatrix} +
$$

$$
+ \begin{bmatrix} h_1(\mathbf{x}) & h_2(\mathbf{x}) & \cdots & h_n(\mathbf{x}) \end{bmatrix}
\begin{bmatrix} \frac{\partial s_1(\mathbf{x})}{\partial x_j} \\ \frac{\partial s_2(\mathbf{x})}{\partial x_j} \\ \vdots \\ \frac{\partial s_n(\mathbf{x})}{\partial x_j} \end{bmatrix}
\tag{2.122}
$$

Notice the order in which we pick the "inner product" (row for $\mathbf{s}$ and column for $\frac{\partial h}{\partial x}$). We could have picked the opposite, resulting in the same summation as in 2.121. The reason we pick that one in particular is because we need to "stack" the result to form a row vector to agree with our definition. Hence, we can write the derivative as

$$\frac{\partial f(\mathbf{x})}{\partial x_j} = \mathbf{s}^t(\mathbf{x}) \begin{bmatrix} \frac{\partial h_1(\mathbf{x})}{\partial x_1} & \frac{\partial h_1(\mathbf{x})}{\partial x_2} & \cdots & \frac{\partial h_1(\mathbf{x})}{\partial x_n} \\ \frac{\partial h_2(\mathbf{x})}{\partial x_1} & \frac{\partial h_2(\mathbf{x})}{\partial x_2} & & \frac{\partial h_2(\mathbf{x})}{\partial x_n} \\ \vdots & & \ddots & \\ \frac{\partial h_m(\mathbf{x})}{\partial x_1} & \frac{\partial h_m(\mathbf{x})}{\partial x_2} & & \frac{\partial h_m(\mathbf{x})}{\partial x_n} \end{bmatrix} +$$

$$+\mathbf{h}^t(\mathbf{x}) \begin{bmatrix} \frac{\partial s_1(\mathbf{x})}{\partial x_1} & \frac{\partial s_1(\mathbf{x})}{\partial x_2} & \cdots & \frac{\partial s_1(\mathbf{x})}{\partial x_n} \\ \frac{\partial s_2(\mathbf{x})}{\partial x_1} & \frac{\partial s_2(\mathbf{x})}{\partial x_2} & & \frac{\partial s_2(\mathbf{x})}{\partial x_n} \\ \vdots & & \ddots & \\ \frac{\partial s_m(\mathbf{x})}{\partial x_1} & \frac{\partial s_m(\mathbf{x})}{\partial x_2} & & \frac{\partial s_m(\mathbf{x})}{\partial x_n} \end{bmatrix} \tag{2.123}$$

Which, using the definition for vector functions differentiation, we can write ■

$$\frac{\partial f(\mathbf{x})}{\partial \mathbf{x}} = \mathbf{s}^t(\mathbf{x})\frac{\partial \mathbf{h}(\mathbf{x})}{\partial \mathbf{x}} + \mathbf{h}^t(\mathbf{x})\frac{\partial \mathbf{s}(\mathbf{x})}{\partial \mathbf{x}} \tag{2.124}$$

2.6.4 Scalar Chain Rule

Let us analyse the chain rule for the composition of scalar functions of one variable with a scalar function of a vector. Formally

$$h(\mathbf{x}) = g(f(\mathbf{x})) \tag{2.125}$$

where the functions are defined as

$$\begin{aligned} f &: \mathfrak{R}^n \to \mathfrak{R} \\ g &: \mathfrak{R} \to \mathfrak{R} \end{aligned} \tag{2.126}$$

We are interested in the following derivative

$$\frac{\partial h(\mathbf{x})}{\partial \mathbf{x}} = \frac{\partial \left(g(f(\mathbf{x}))\right)}{\partial \mathbf{x}} \tag{2.127}$$

As usual, we differentiate with respect to each component as

$$\frac{\partial h(\mathbf{x})}{\partial x_i} = \frac{\partial \left(g(f(\mathbf{x})) \right)}{\partial x_i}$$
$$= \frac{\partial g(f)}{\partial f} \frac{\partial f(\mathbf{x})}{\partial x_i} \tag{2.128}$$

and assemble the resulting vector as ■

$$\frac{\partial h(\mathbf{x})}{\partial \mathbf{x}} = \frac{\partial g(f)}{\partial f} \frac{\partial f(\mathbf{x})}{\partial \mathbf{x}} \tag{2.129}$$

Notice that the first differentiation is done with one variable in the normal scalar-scalar case, hence the order of multiplication doesn't matter.

2.6.5   Chain Rule For Vector Functions

Now, we focus on the chain rule for vector-valued functions. We are interested in the following case

$$h(\mathbf{x}) = g(\mathbf{f}(\mathbf{x})) \tag{2.130}$$

with

$$f : \Re^n \to \Re^m$$
$$g : \Re^m \to \Re \tag{2.131}$$

The differentiation we will analyse is

$$\frac{\partial h(\mathbf{x})}{\partial \mathbf{x}} = \frac{\partial \left(g(\mathbf{f}(\mathbf{x})) \right)}{\partial \mathbf{x}} \tag{2.132}$$

Differentiating with respect to each component of $\mathbf{x}$, we have

$$\frac{\partial h(\mathbf{x})}{\partial x_1} = \frac{\partial \left(g(\mathbf{f}(\mathbf{x})) \right)}{\partial x_1} \tag{2.133}$$

Now we have a slightly different situation. We are differentiating a scalar function of several variables that themselves depend of the the differential variable (x_i). We need to apply the chain rule for a function of several variables as in section 2.2.2. Hence, we have

$$\frac{\partial \left(g(\mathbf{f}(\mathbf{x}))\right)}{\partial x_1} = \frac{\partial \left(g(\mathbf{f}(\mathbf{x}))\right)}{\partial f_1}\frac{\partial f_1(\mathbf{x})}{\partial x_i} + \frac{\partial \left(g(\mathbf{f}(\mathbf{x}))\right)}{\partial f_2}\frac{\partial f_2(\mathbf{x})}{\partial x_i} + \cdots +$$
$$+\frac{\partial \left(g(\mathbf{f}(\mathbf{x}))\right)}{\partial f_m}\frac{\partial f_m(\mathbf{x})}{\partial x_i}$$

$$(2.134)$$

which can be written as an inner product as

$$\frac{\partial \left(g(\mathbf{f}(\mathbf{x}))\right)}{\partial x_1} = \mathbf{u}^t\mathbf{v} \qquad (2.135)$$

where

$$\mathbf{u}^t = \frac{\partial \left(g(\mathbf{f}(\mathbf{x}))\right)}{\partial \mathbf{f}}$$
$$\mathbf{v} = \frac{\partial \mathbf{f}(\mathbf{x})}{\partial x_i} \qquad (2.136)$$

Now, if we assemble the vector for each component, we have

$$\frac{\partial h(\mathbf{x})}{\partial \mathbf{x}} = \begin{bmatrix} \mathbf{u}^t\mathbf{v}_1 & \mathbf{u}^t\mathbf{v}_2 & \cdots & \mathbf{u}^t\mathbf{v}_n \end{bmatrix}$$
$$= \mathbf{u}^t\mathbf{V} \qquad (2.137)$$

where

$$\mathbf{V} = \begin{bmatrix} \mathbf{v}_1 & \mathbf{v}_2 & \cdots & \mathbf{v}_n \end{bmatrix}$$
$$= \frac{\partial \mathbf{f}(\mathbf{x})}{\partial \mathbf{x}} \qquad (2.138)$$

That finally leads us to ■

$$\frac{\partial \left(g(\mathbf{f}(\mathbf{x}))\right)}{\partial \mathbf{x}} = \frac{\partial \left(g(\mathbf{f}(\mathbf{x}))\right)}{\partial \mathbf{f}}\frac{\partial \mathbf{f}(\mathbf{x})}{\partial \mathbf{x}} \qquad (2.139)$$

Notice that as trivial as it might seem, this result is derived from basic principles and, as the second term is a matrix, the order of multiplication is important.

2.6.6 Vector norm

We now will derive the derivative of the vector norm with respect to the vector itself. We want the following derivative

$$f(\mathbf{x}) = \frac{\partial \|\mathbf{x}\|}{\partial \mathbf{x}} \tag{2.140}$$

We will use the common norm definition given by

$$\|\mathbf{x}\| = \sqrt{\sum_i x_i^2} \tag{2.141}$$

Differentiating with respect to each component we get

$$\frac{\partial \|\mathbf{x}\|}{\partial x_i} = \frac{x_i}{\sqrt{\sum_i x_i^2}} = \frac{x_i}{\|\mathbf{x}\|} \tag{2.142}$$

and assembling the final vector form, we have ■

$$\frac{\partial \|\mathbf{x}\|}{\partial \mathbf{x}} = \frac{\mathbf{x}^t}{\|\mathbf{x}\|} \tag{2.143}$$

Notice the transposition of $\mathbf{x}$, indicating that the result must comply with the numerator layout for the result.

2.6.7   Linear form

One very common and useful vector function is the form

$$\mathbf{f}(\mathbf{x}) = \mathbf{A}\mathbf{x} \tag{2.144}$$

Before differentiating, we will expand the product as a set of inner products between the rows of $\mathbf{A}$ and $\mathbf{x}$ as

$$\mathbf{f}(\mathbf{x}) = \begin{bmatrix} \mathbf{a}_1^t \mathbf{x} \\ \mathbf{a}_2^t \mathbf{x} \\ \vdots \\ \mathbf{a}_m^t \mathbf{x} \end{bmatrix} \tag{2.145}$$

where

$$\mathbf{a}_i = \begin{bmatrix} a_{i,1} \\ a_{i,2} \\ \vdots \\ a_{i,m} \end{bmatrix} \tag{2.146}$$

Now, we can proceed as usual and write the derivative as (see 2.6.2)

$$\frac{\partial \mathbf{f}(\mathbf{x})}{\partial \mathbf{x}} = \begin{bmatrix} \frac{\partial \mathbf{a}_1^t \mathbf{x}}{\partial \mathbf{x}} \\ \frac{\partial \mathbf{a}_2^t \mathbf{x}}{\partial \mathbf{x}} \\ \vdots \\ \frac{\partial \mathbf{a}_m^t \mathbf{x}}{\partial \mathbf{x}} \end{bmatrix} = \begin{bmatrix} \mathbf{a}_1^t \\ \mathbf{a}_2^t \\ \vdots \\ \mathbf{a}_m^t \end{bmatrix} \qquad (2.147)$$

which is simply ■

$$\frac{\partial \mathbf{f}(\mathbf{x})}{\partial \mathbf{x}} = \mathbf{A} \qquad (2.148)$$

Transpose form

If we have the linear form like

$$\mathbf{f}(\mathbf{x}) = \mathbf{x}^t \mathbf{A} \qquad (2.149)$$

we would have the following form for $\mathbf{f}(\mathbf{x})$

$$\mathbf{f}(\mathbf{x}) = \begin{bmatrix} \mathbf{x}^t \mathbf{a}'_1 \\ \mathbf{x}^t \mathbf{a}'_2 \\ \vdots \\ \mathbf{x}^t \mathbf{a}'_m \end{bmatrix} \qquad (2.150)$$

where

$$\mathbf{a}'_i = \begin{bmatrix} a_{1,i} \\ a_{2,i} \\ \vdots \\ a_{n,i} \end{bmatrix} \qquad (2.151)$$

which differentiating results in

$$\frac{\partial \mathbf{f}(\mathbf{x})}{\partial \mathbf{x}} = \begin{bmatrix} \frac{\partial \mathbf{x}^t \mathbf{a}'_1}{\partial \mathbf{x}} \\ \frac{\partial \mathbf{x}^t \mathbf{a}'_2}{\partial \mathbf{x}} \\ \vdots \\ \frac{\partial \mathbf{x}^t \mathbf{a}'_3}{\partial \mathbf{x}} \end{bmatrix} = \begin{bmatrix} \mathbf{a}'^t_1 \\ \mathbf{a}'^t_2 \\ \vdots \\ \mathbf{a}'^t_m \end{bmatrix} \qquad (2.152)$$

which is simply ■

$$\frac{\partial \mathbf{f}(\mathbf{x})}{\partial \mathbf{x}} = \mathbf{A}^t \tag{2.153}$$

2.6.8   Quadratic form

Another very useful form for vector expressions is the quadratic form

$$f(\mathbf{x}) = \mathbf{x}^t \mathbf{A} \mathbf{x} \tag{2.154}$$

We will differentiate by first writing it as the product

$$f(\mathbf{x}) = \mathbf{x}^t \left(\mathbf{A} \mathbf{x} \right) \tag{2.155}$$

Using the result of section 2.6.5, we write

$$\frac{\partial f(\mathbf{x})}{\partial \mathbf{x}} = (\mathbf{A}\mathbf{x})^t \frac{\partial \mathbf{x}}{\partial \mathbf{x}} + \mathbf{x}^t \frac{\partial \mathbf{A}\mathbf{x}}{\partial \mathbf{x}} \tag{2.156}$$

and use the linear form in the previous sections to get [1]

$$\frac{\partial f(\mathbf{x})}{\partial \mathbf{x}} = (\mathbf{A}\mathbf{x})^t + \mathbf{x}^t \mathbf{A} \tag{2.157}$$

and finally have ■

$$\frac{\partial f(\mathbf{x})}{\partial \mathbf{x}} = \mathbf{x}^t \left(\mathbf{A} + \mathbf{A}^t \right) \tag{2.158}$$

---

[1] We use the fact that we can write a vector as itself times the identity matrix $\mathbf{x} = \mathbf{I}\mathbf{x}$

Chapter 3

Probability and Statistics

N ext, we are going to present a selection of derivations for probability and statistics subjects. The reader must be familiar with the basics of probability, but most of the derivations will be accompanied by a simple explanatory text. It is important to keep in mind that the derivations are not an introduction to the subject, even though some detail of the theory is given. We will assume a basic prior knowledge by the reader.

3.1 Gaussian Integral

Consider the Gaussian distribution

$$p(x) = \frac{1}{\sqrt{2\pi\sigma^2}} e^{-\frac{(x-\mu)^2}{2\sigma^2}} \qquad (3.1)$$

where μ and σ^2 are the mean and variance respectively of the RV (random variable). One of the best known facts is that the integral is equal to one (since it is a pdf - probability density function). In this section we will compute the integral and show how we get to this result. Probably the easiest way to solve the integral is to write the square of it in terms of another integral with a different variable. So, let us focus on the integral only and write it as a simple positive value I as $\sqrt{I^2}$. Doing that we have

$$\sqrt{\int_{-\infty}^{\infty} e^{-\frac{x^2}{2\sigma^2}}\, dx \int_{-\infty}^{\infty} e^{-\frac{y^2}{2\sigma^2}}\, dy}. \tag{3.2}$$

As the functions are separable, we can group them in the following way

$$I = \sqrt{\int_{-\infty}^{\infty}\int_{-\infty}^{\infty} e^{-\frac{y^2}{2\sigma^2}} e^{-\frac{x^2}{2\sigma^2}}\, dy\, dx} \tag{3.3}$$

that we can rewrite as

$$I = \sqrt{\int_{-\infty}^{\infty}\int_{-\infty}^{\infty} e^{-\frac{1}{2\sigma^2}(x^2+y^2)}\, dy\, dx}. \tag{3.4}$$

Now, we can perform the following change of variables

$$x^2 + y^2 = r^2 \tag{3.5}$$

and

$$dx\, dy = r\, dr\, d\theta \tag{3.6}$$

This change of variables corresponds to integration in all of the $\mathbb{R}^2$, but in polar coordinates. This procedure leads to the following integral

$$I = \sqrt{\int_{0}^{2\pi}\int_{0}^{\infty} r e^{-\frac{1}{2\sigma^2}r^2}\, dr\, d\theta} \tag{3.7}$$

Therefore, we can now compute the integral as

$$I = \sqrt{2\pi}\sigma$$

which is the denominator of equation 3.1 ∎.

3.2 Maximum Likelihood estimation

A very practical problem in probability and statistics is the estimation of parameters for some distribution. That might sound a very technical question, but in fact it is one of the most intuitive and yet non-trivial ones to answer. For instance, we know that for a Gaussian distribution, the mean is the centre of its distribution and the variance is the "width". But, what if we only have samples from that distribution? That is in general exactly what we have in practice. No one has the "distribution" for the ages in a classroom, or the size of a population. What we have are measurements or samples. So, the question that arises is: Assuming a Gaussian distributed variable, how do we compute the mean and variance, given a set of samples?

To answer that question, we need a theory called parameter estimation and we will, in this section, use a principle called Maximum Likelihood (ML) to estimate the mean and variance (from a Gaussian) from a set of samples. The Maximum Likelihood estimation consists of finding the parameter Θ for a distribution $p(x|\Theta)$, given the set of measurements $X = \{x_1, x_2, \ldots, x_N\}$. We do that by maximising the function

$$L(\Theta) = \prod_{i=1}^{N} p(x_i|\Theta) \qquad (3.8)$$

where $p(x_i|\Theta)$ is the probability of the measurement x_i happening, given the parameter value Θ.

The rationale behind this functional is that it measures the probability of all the points in this set of measurements happening (considering that they are made independent of each other). So, since those are the measurements that actually happened, it makes sense that they have a great probability of happening. Hence, we choose the parameter Θ that maximises $L(\Theta)$.

Normally, it is easier to maximise the logarithm of the likelihood $L(\theta)$, which gives the same optimum. So, the function becomes ■

$$l(\Theta) = \sum_{i=1}^{N} \log(p(x_i|\Theta)) \tag{3.9}$$

3.2.1   Mean of a Gaussian

Now, we will apply the ML principle to a Gaussian with mean μ and variance σ^2. We will first optimise the likelihood to find the mean of the Gaussian given some samples. Our parameter will be the mean $\Theta = \mu$ and the density is given by

$$P(x|\mu) = \frac{1}{\sqrt{2\pi\sigma^2}} e^{-\frac{1}{2\sigma^2}(x-\mu)^2} \tag{3.10}$$

Plugging that into the log-likelihood expression, we have

$$l(\Theta) = \sum_{i=1}^{N} \log\left(\frac{1}{\sqrt{2\pi\sigma^2}} e^{-\frac{1}{2\sigma^2}(x_i-\mu)^2}\right) \tag{3.11}$$

Expanding the log and separating the sum, we have

$$l(\Theta) = \sum_{i=1}^{N} -0.5\log(2\pi\sigma^2) - \frac{1}{2\sigma^2}(x_i - \mu)^2 \tag{3.12}$$

Now, differentiating with respect to μ, we have

$$\frac{\partial l(\Theta)}{\partial \mu} = -\sum_{i=1}^{N} \frac{1}{\sigma^2}(x_i - \mu) \tag{3.13}$$

Now, we can equate to zero and solve for μ and get

$$0 = -\frac{1}{\sigma^2}\sum_{i=1}^{N}(x_i - \mu)$$

$$= \sum_{i=1}^{N}(x_i - \mu) \tag{3.14}$$

which finally leads to ∎

$$\mu = \frac{1}{N} \sum_{i=1}^{N} x_i \qquad (3.15)$$

It is remarkable that we use this simple formula to compute the mean of a population of measurements, sometimes without realising the non-trivial origin of it. In the next section we will do the same for the variance.

3.2.2 Variance of a Gaussian

Now, we start with the same Gaussian form for the function as in eq. 3.11, but now we differentiate with respect to its variance. So we now have

$$P(x|\sigma^2) = \frac{1}{\sqrt{2\pi\sigma^2}} e^{-\frac{1}{2\sigma^2}(x-\mu)^2} \qquad (3.16)$$

and

$$l(\Theta) = \sum_{i=1}^{N} -0.5 \log(2\pi\sigma^2) - \frac{1}{2\sigma^2}(x_i - \mu)^2 \qquad (3.17)$$

Differentiating with respect to. σ^2, we have

$$\frac{\partial l(\Theta)}{\partial \sigma} = \sum_{i=1}^{N} -\frac{0.5}{\sigma^2} + \frac{0.5}{\sigma^4}(x_i - \mu)^2 \qquad (3.18)$$

Now, equating to zero and solving the σ^2, we have

$$0 = \sum_{i=1}^{N} -\frac{0.5}{\sigma^2} + \frac{0.5}{\sigma^4}(x_i - \mu)^2$$
$$= \sum_{i=1}^{N} -\sigma^2 + (x_i - \mu)^2$$

$\qquad\qquad (3.19)$

and finally ■

$$\sigma^2 = \frac{1}{N} \sum_{i=1}^{N} (x_i - \mu)^2 \qquad (3.20)$$

.

3.3   Maximum Entropy Principle

In this section we are going to visit the problem of the Maximum Entropy Principle (MEP). As this is an important subject in the probability area, we will try to give some level of textual detail to the derivation. We will divide the derivation for the MEP into its discrete and continuous cases in the sections ahead.

"In the absence of prior information, the distribution that models some random variable must be uniformly distributed."

3.3.1   Discrete case

We will derive two versions of the MEP. In this section we will derive the discrete version and in the next section we will derive the continuous version. Let x be a random variable (RV) assuming integer values between 0 and N. The probability of x being equal to some i is denoted by $p_x[i]$. In the Maximum Entropy Principle we have the concept of prior information. Although one can define prior information in several ways, we will formalise it by means of measures in the random variable. We then define a "piece of information" in the random variable x as any measure of any of its moments or probabilities directly. For instance, one can say: "I have a RV whose mean is 3.14" or say "I have a RV where the probability of $x = 2$ is 0.5". In another words, one can have any measurable information about the variable.

It will be taken as a given, that any variable MUST obey the MEP, as it relates the distribution of some RV to probability values that model all possible information about that variable.

If one takes that last statement as a definition, we must take MEP as a definition too. There are several ways to measure information (depending on the unity one wants to measure). Using bits as unity, Shannon Entropy can be stated as (for a discrete RV)

$$H(x) = \sum_{i=0}^{N} p_x[i] \, \log_2(p_x[i]) \tag{3.21}$$

We now want to maximise that entropy with respect to the probabilities, given all the information we have. To do that, our goal is to write a functional that incorporates the entropy and restricts the probabilities to ones that meet all the M restrictions imposed by knowing the information we have a priori. In another words, we want to solve the following problem

$$\begin{cases} \min_{p_x[i]} H(x) \\ f_j(x) = m_j, \quad j = 0, 1, \ldots, M \end{cases} \tag{3.22}$$

The function that one must use to solve this problem can be written as

$$J(x) = H(x) + \sum_{j=0}^{M} \lambda_j \left(f_j(x) - m_j \right) \tag{3.23}$$

where the second term on the right-hand side of the expression is the Lagrangian term for the restrictions. λs are the Lagrangian multipliers, $f_j(x)$ are the expressions for the measurements and m_j are the measurement values (information that is given a priori).

For instance, if one knows the mean m_1 of the RV, the Lagrangian multiplier term is written as

$$\lambda_1 \left(\sum_{i=0}^{N} i p_x[i] - m_1 \right) \tag{3.24}$$

The fundamental restriction (normally attributed to the first multiplier $j = 0$) for $p_x[i]$ being a probability distribution can be given as (sum of all probabilities equals to one)

$$\lambda_0 \left(\sum_{i=0}^{N} p_x[i] - 1 \right) \tag{3.25}$$

There are several ways to achieve the goal of optimising this function, and some involve variational calculus, or analysis. We will take an algebraic approach, for it is more intuitive.

First, as we will focus on the case where we have no a priori information, the function can be written as

$$J(x) = \sum_{i=0}^{N} p_x[i] \log_2 (p_x[i]) + \lambda_0 \left(\sum_{i=0}^{N} p_x[i] - 1 \right) \tag{3.26}$$

meaning that we want the distribution that models ALL the possible information about the RV (so we take the MEP as a definition) and impose ONLY that this distribution is a probability distribution (sum up to one). No other information is available (so we have no Lagrangian multipliers besides λ_0).

Let us consider a small increment $\delta p_x[i]$ in the probability distribution in any direction. So, the new probability distribution is $p_x[i] + \delta p_x[i]$. The new function also increases by a small amount $\delta J(x)$ and becomes

$$J(x) + \delta J(x) = \sum_{i=0}^{N} (p_x[i] + \delta p_x[i]) \log_2 (p_x[i] + \delta p_x[i])$$

$$+ \lambda_0 \left(\sum_{i=0}^{N} p_x[i] + \delta p_x[i] - 1 \right) \tag{3.27}$$

Now, we develop a little and expand the new function to

$$J(x) + \delta J(x) =$$

$$-1 + \sum_{i=0}^{N} p_x[i]\log_2 (p_x[i] + \delta p_x[i]) + \delta p_x[i]\log_2 (p_x[i] + \delta p_x[i]) +$$

$$+\lambda_0 p_x[i] + \lambda_0 \delta p_x[i]$$

$$(3.28)$$

We notice then that, as $\delta p_x[i]$ tends to be smaller (towards zero), we can write the log as the first term of its expansion $(\log_2(A + a) = \log_2(A) + \frac{a}{A})$ and write

$$J(x) + \delta J(x) =$$

$$-1 + \sum_{i=0}^{N} p_x[i]\log_2 (p_x[i]) + p_x[i]\frac{\delta p_x[i]}{p_x[i]} + \delta p_x[i]\log_2 (p_x[i]) + \delta p_x[i]\frac{\delta p_x[i]}{p_x[i]} +$$

$$+\lambda_0 p_x[i] + \lambda_0 \delta p_x[i]$$

$$(3.29)$$

giving (neglecting the squared term that will approach zero quicker than the others)

$$J(x) + \delta J(x) =$$

$$-1 + \sum_{i=0}^{N} p_x[i]\log_2 (p_x[i]) + \delta p_x[i] + \delta p_x[i]\log_2 (p_x[i]) + \lambda_0 p_x[i] + \lambda_0 \delta p_x[i]$$

$$(3.30)$$

Now, we identify that we can cancel out the term $J(x)$ on the left-hand side with the corresponding terms on the right-hand side, leaving only

$$\delta J(x) = \sum_{i=0}^{N} \delta p_x[i] \left(1 + \log_2 (p_x[i]) + \lambda_0\right) \qquad (3.31)$$

The reasoning now is that we want the optimum point for $J(x)$. By analogy with maximising a function, the maximum occurs where the variation of the argument causes no variation in the value of the function (commonly stated as "differentiate and equate to zero"). In this case, we need to make $\delta J(x) = 0$ regardless of $\delta p_x[i]$. This is possible only by making the term in parentheses in the last equation equal to zero. And that leads us to

$$1 + \log_2 (p_x[i]) + \lambda_0 = 0 \qquad (3.32)$$

Finally, solving for $p_x[i]$, we have ■

$$p_x[i] = 2^{-1-\lambda_0} \qquad (3.33)$$

The Lagrangian multiplier must be chosen to meet the restrictions (that is, making the sum up to one).

As one can see, having no prior information in the MEP derivation, it leads us to a distribution where all the probabilities are equal, proving algebraically the principle for the uniform distribution.

3.3.2 Continuous case

Now, we will derive the same conclusion as for the MEP in the discrete case, but now we will assume a continuous random variable x with density $p(x)$. We will also use Shannon's entropy as the information measure and will consider a finite support distribution.

As we did in the discrete case, let us define the function (including the Lagrangian term for the restriction)

$$J(p(x)) = H(p(x)) + \lambda \left(1 - \int_a^b p(x)\mathrm{d}x \right) \qquad (3.34)$$

Using Shannon's definition for entropy, we have

$$J(p(x)) = - \int_a^b p(x) \log (p(x))\, \mathrm{d}x + \lambda \left(1 - \int_a^b p(x)\mathrm{d}x \right) \qquad (3.35)$$

Using the differential approach from functional calculus and performing an increment $\delta p(x)$ to $p(x)$, we have

$$J(p(x) + \delta p(x)) = -\int_a^b (p(x) + \delta p(x)) \log (p(x) + \delta p(x))\, dx +$$

$$\lambda \left(1 - \int_a^b (p(x) + \delta p(x))\, dx \right)$$

$$(3.36)$$

Isolating $\delta p(x)$ and approximating for small changes, we have

$$J(p(x)) + \delta J(p(x)) = -\int_a^b (p(x) + \delta p(x)) \left(\log (p(x)) + \frac{\delta p(x)}{p(x)} \right) dx +$$

$$\lambda \left(1 - \int_a^b p(x)\, dx - \int_a^b \delta p(x)\, dx \right)$$

$$= -\int_a^b \left(p(x) \left(\log (p(x)) + \frac{\delta p(x)}{p(x)} \right) + \delta p(x) \left(\log (p(x)) + \frac{\delta p(x)}{p(x)} \right) \right) dx +$$

$$\lambda \left(1 - \int_a^b p(x)\, dx - \int_a^b \delta p(x)\, dx \right)$$

$$(3.37)$$

With some more algebra, we have

$$J(p(x)) + \delta J(p(x)) =$$

$$-\int_a^b \left(p(x)\log{(p(x))} + \delta p(x) + \delta p(x)\log{(p(x))} + \frac{\delta p^2(x)}{p(x)} - \lambda \delta p(x) \right) dx$$

$$= -\int_a^b p(x)\log{(p(x))}\, dx$$

$$-\int_a^b \left(\delta p(x) + \delta p(x)\log{(p(x))} + \frac{\delta p^2(x)}{p(x)} - \lambda \delta p(x) \right) dx$$

$$(3.38)$$

Now, we eliminate the term $J(p(x))$ from both sides and have

$$\delta J(p(x)) = -\int_a^b \left(\delta p(x) + \delta p(x)\log{(p(x))} + \frac{\delta p^2(x)}{p(x)} - \lambda \delta p(x) \right) dx$$

$$(3.39)$$

As the small change in $p(x)$ ($\delta p(x)$) tends to zero, the quadratic term vanishes and we can write

$$\delta J(p(x)) = -\int_a^b \left(\delta p(x) + \delta p(x)\log{(p(x))} - \lambda \delta p(x) \right) dx \quad (3.40)$$

Now, we have to make the differential $\delta J(p(x))$ equal to zero (to find an optimum of the function). Since it is inside an integral, the term that multiplies the differential $\delta p(x)$ must be zero to ensure that the integral is zero. Hence we have

$$\int_a^b \delta p(x)\left(1 + \log{(p(x))} - \lambda\right) dx = 0 \qquad (3.41)$$

That leads us to

$$1 + \log\left(p(x)\right) - \lambda = 0$$
$$p(x) = e^{\lambda - 1} \tag{3.42}$$

which states that $p(x)$ must be constant in the interval (a, b) ■.

3.4 Entropy of a Gaussian

In the following sections we will derive the expressions for the entropy of a random variable that is Gaussian distributed. That entropy normally is presented without derivation because it is the result only of an integral calculation. We will then provide that calculation.

3.4.1 Shannon Entropy

Here we will derive the Shannon entropy for a multivariate Gaussian (general case). The mono-variable case results in an integral that is practically the same as the Gaussian integral presented in section 3.1.

We will start with the expression for the pdf of a multivariable Gaussian random variable as in the following equation

$$p(\mathbf{x}) = \frac{1}{\sqrt{|2\pi\,\mathbf{\Sigma}|}} e^{-\frac{1}{2}(\mathbf{x}-\mu)^t \mathbf{\Sigma}^{-1}(\mathbf{x}-\mu)} \tag{3.43}$$

where μ is the mean of the RV and $\mathbf{\Sigma}$ is its covariance matrix.

The Shannon entropy is defined as

$$H\left(p(\mathbf{x})\right) = -\int_{-\infty}^{\infty} \cdots \int_{-\infty}^{\infty} p(\mathbf{x}) \log\left(p(\mathbf{x})\right) d\mathbf{x} \tag{3.44}$$

Substituting the expression of the Gaussian pdf into it, we have (after some simple algebraic manipulations)

$$H\left(p(\mathbf{x})\right) = \frac{\log\left(|2\pi\boldsymbol{\Sigma}|\right)}{2} +$$

$$\int\limits_{-\infty}^{\infty}\cdots\int\limits_{-\infty}^{\infty}\frac{1}{2\sqrt{|2\pi\boldsymbol{\Sigma}|}}e^{-\frac{1}{2}(\mathbf{x}-\mu)^t\boldsymbol{\Sigma}^{-1}(\mathbf{x}-\mu)}(\mathbf{x}-\mu)^t\boldsymbol{\Sigma}^{-1}\left(\mathbf{x}-\mu\right)\mathrm{d}\mathbf{x}$$

$$(3.45)$$

Now, we make a simple change of variable $\mathbf{y} = \mathbf{x} - \mu$, which does not change the limits of integration and we get

$$H\left(p(\mathbf{x})\right) = \frac{\log\left(|2\pi\boldsymbol{\Sigma}|\right)}{2}$$

$$+\frac{1}{2\sqrt{|2\pi\boldsymbol{\Sigma}|}}\int\limits_{-\infty}^{\infty}\cdots\int\limits_{-\infty}^{\infty}e^{-\frac{1}{2}\mathbf{y}^t\boldsymbol{\Sigma}^{-1}\mathbf{y}}\mathbf{y}^t\boldsymbol{\Sigma}^{-1}\mathbf{y}\,\mathrm{d}\mathbf{y} \qquad (3.46)$$

We now have a very mixed set of variables that makes the multiple integral difficult to solve. So, the idea now is that, since we are interested in the hyper-volume of a multidimensional function (portion of the formula within the integral), it doesn't matter if we rotate the whole volume. So, we proceed with a new change of variables that produces a rotation in space. In a multidimensional Euclidian space, a rotation can be described by a multiplication to a unitary matrix. Using this fact, we choose the unitary matrix that turns the covariance matrix $\boldsymbol{\Sigma}$ into a diagonal matrix (remember that as a symmetric matrix, the covariance matrix can always be diagonalised). We then make

$$\boldsymbol{\Sigma}^{-1} = \mathbf{U}^t\boldsymbol{\Lambda}^{-1}\mathbf{U}$$
$$\mathbf{z} = \mathbf{U}\mathbf{y} \qquad (3.47)$$

and defining I as the integral we are computing, we have

$$I = \int\limits_{-\infty}^{\infty}\cdots\int\limits_{-\infty}^{\infty}e^{-\frac{1}{2}\mathbf{z}^t\boldsymbol{\Lambda}^{-1}\mathbf{z}}\mathbf{z}^t\boldsymbol{\Lambda}^{-1}\mathbf{z}\,|\mathbf{U}|\,\mathrm{d}\mathbf{z} \qquad (3.48)$$

Now, as Λ is diagonal and U is unitary (determinant equals 1), we can write

$$I = \int_{-\infty}^{\infty} \dots \int_{-\infty}^{\infty} e^{-\frac{1}{2}\sum_i \frac{1}{\lambda_i} z_i^2} \sum_i \frac{1}{\lambda_i} z_i^2 dz_1 \dots dz_n \qquad (3.49)$$

and exchange the summation as

$$I = \sum_j \frac{1}{\lambda_j} \int_{-\infty}^{\infty} \dots \int_{-\infty}^{\infty} e^{-\frac{1}{2}\sum_i \frac{1}{\lambda_i} z_i^2} z_j^2 dz_1 \dots dz_n \qquad (3.50)$$

Now, the only thing that prevents us having several separated Gaussian integrals is the z_j term multiplying the exponential. So, we isolate the jth term in the exponential and divide the integral into two parts as

$$I = \sum_j \frac{1}{\lambda_j} \int_{-\infty}^{\infty} \dots \int_{-\infty}^{\infty} e^{-\frac{1}{2\lambda_j} z_j^2} e^{-\frac{1}{2}\sum_{i\neq j} \frac{1}{\lambda_i} z_i^2} z_j^2 dz_1 \dots dz_n \qquad (3.51)$$

$$I = \sum_j \frac{1}{\lambda_j} \int_{-\infty}^{\infty} \dots \int_{-\infty}^{\infty} e^{-\frac{1}{2}\sum_{i\neq j} \frac{1}{\lambda_i} z_i^2} dz_1 \dots dz_{j-1} dz_{j+1} \dots dz_n\, A$$

$$A = \int_{-\infty}^{\infty} e^{-\frac{1}{2\lambda_j} z_j^2} z_j^2 dz_j$$

$$(3.52)$$

and write a compact product notation as

$$I = \sum_j \frac{1}{\lambda_j} \int_{-\infty}^{\infty} e^{-\frac{z_j^2}{2\lambda_j}} z_j^2 dz_j \prod_{i\neq j} \int_{-\infty}^{\infty} e^{-\frac{z_i^2}{2\lambda_i}} dz_i \qquad (3.53)$$

We can now identify the separated integrals and solve (after which we apply some algebraic simplifications)

$$I = \sum_j \frac{1}{\lambda_j} \int_{-\infty}^{\infty} e^{-\frac{z_j^2}{2\lambda_j}} z_j^2 dz_j \prod_{i \neq j} \sqrt{2\pi\lambda_i} \qquad (3.54)$$

$$I = \sum_j \sqrt{2\pi\lambda_j} \prod_{i \neq j} \sqrt{2\pi\lambda_i} \qquad (3.55)$$

$$I = n\sqrt{|2\pi\Sigma|} \qquad (3.56)$$

Going back and substituting the expression for the integral, we have

$$H\left(p(\mathbf{x})\right) = \frac{\log\left(|2\pi\Sigma|\right)}{2} + \frac{n}{2} \qquad (3.57)$$

That, after a somewhat elegant algebraic simplification, leads us to

$$H\left(p(\mathbf{x})\right) = \frac{1}{2} \log\left(|2\pi e\Sigma|\right) \qquad (3.58)$$

which is the formula that is generally presented in the books ■.

3.4.2   Rényi Entropy

Now, we will derive the Rényi version of the previous derivation. Rényi's entropy of an RV x with com density $p(\mathbf{x})$ is given by

$$H_\alpha = \frac{1}{1-\alpha} \log \left(\int_{-\infty}^{\infty} \cdots \int_{-\infty}^{\infty} p(\boldsymbol{x})^\alpha d\boldsymbol{x} \right) \qquad (3.59)$$

For a gaussian RV the expression becomes

$$H_\alpha = \frac{1}{1-\alpha} \log \left(\int_{-\infty}^{\infty} \cdots \int_{-\infty}^{\infty} \left(k e^{-\frac{1}{2}(\boldsymbol{x}-\boldsymbol{\mu})^t \boldsymbol{\Sigma}^{-1}(\boldsymbol{x}-\boldsymbol{\mu})} \right)^\alpha d\boldsymbol{x} \right) \qquad (3.60)$$

where k is a constant equal to

$$k = \frac{1}{\sqrt{|2\pi\Sigma|}}. \tag{3.61}$$

Performing a simple change of variables as

$$y = x - \mu \tag{3.62}$$

we obtain

$$H_\alpha = \frac{1}{1-\alpha} \log\left(k^\alpha \int_{-\infty}^{\infty} \cdots \int_{-\infty}^{\infty} e^{-\frac{\alpha}{2}y^t \Sigma^{-1} y} dy\right). \tag{3.63}$$

Now, doing

$$\Sigma = \alpha\Sigma_q, \tag{3.64}$$

and substituting, we have

$$H_\alpha = \frac{1}{1-\alpha} \log\left(k^\alpha \int_{-\infty}^{\infty} \cdots \int_{-\infty}^{\infty} e^{-\frac{1}{2}y^t \Sigma_q^{-1} y} dy\right) \tag{3.65}$$

Solving the integral (as shown in the previous sections), we have

$$H_\alpha = \frac{1}{1-\alpha} \log\left(k^\alpha \sqrt{|2\pi\Sigma_q|}\right) \tag{3.66}$$

$$H_\alpha = \frac{1}{1-\alpha} \log\left(k^\alpha \sqrt{\alpha^{-D} |2\pi\Sigma|}\right) \tag{3.67}$$

Now, putting back the value for k, we get

$$H_\alpha = \frac{1}{1-\alpha} \log\left(\frac{\sqrt{\alpha^{-D} |2\pi\Sigma|}}{|2\pi\Sigma|^{\frac{\alpha}{2}}}\right) \tag{3.68}$$

That simplifies to

$$H_\alpha = \frac{1}{1-\alpha} \log\left(\alpha^{-D} |2\pi\mathbf{\Sigma}|^{\frac{1}{2}(1-\alpha)}\right)$$
$$H_\alpha = \frac{1}{2} \log\left(\alpha^{-\frac{D}{1-\alpha}} |2\pi\mathbf{\Sigma}|\right) \qquad (3.69)$$
$$H_\alpha = \frac{1}{2} \log\left(\left|\alpha^{\frac{1}{\alpha-1}} 2\pi\mathbf{\Sigma}\right|\right)$$

For the case of quadratic entropy, we get

$$H_2 = \frac{1}{2} \log\left(|4\pi\mathbf{\Sigma}|\right), \qquad (3.70)$$

For the case of $\alpha = 1$, we should obtain the Shannon entropy (as it is known in the literature). We just need to notice that the term $\alpha^{\frac{1}{\alpha-1}}$ when $\alpha \to 1$ is one of the limits for Euler's Number e. So, we have ■

$$H_1 = \frac{1}{2} \log\left(|2\pi\, e\mathbf{\Sigma}|\right) \qquad (3.71)$$

3.5   Linear Fischer Discriminant

In binary classification problems, the goal is to classify an n-dimensional data point (representing measurements in a RV or process) into one of two possible classes. In the Fischer Discriminant analysis, that is done by modelling each class as a Gaussian with means μ_1 and μ_2 and covariance matrices $\mathbf{\Sigma}_1$ and $\mathbf{\Sigma}_2$. The probabilistic approach to decide whether a data point x belongs to one class or another is to compute the probability of the point belonging to each class as

$$\begin{aligned} p(\mathbf{x}|C_1) \\ p(\mathbf{x}|C_2) \end{aligned} \qquad (3.72)$$

where $p(.)$ is the probability function and C_1 and C_2 are the indicators of the class. Since in the Fischer's paradigm we have Gaussians, one can define the parameters for each class as

$$C_1 : \mu_1, \Sigma_1$$
$$C_2 : \mu_2, \Sigma_2$$
(3.73)

and therefore the probabilities will be given by

$$p(x|C_i) = \frac{P_i}{\sqrt{|2\pi\Sigma_i|}} e^{-\frac{1}{2}(\mathbf{x}-\mu_i)^t \Sigma_i^{-1}(\mathbf{x}-\mu_i)}$$
(3.74)

Now, it is possible to assemble an hypothesis test for class 1, say, as

$$H_0 : x \in C_1 \ if; p(x|C_1) > p(x|C_2)$$
(3.75)

that will be formally tested as

$$\frac{P_1}{\sqrt{|2\pi\Sigma_1|}} e^{-\frac{1}{2}(x-\mu_1)^t \Sigma_1^{-1}(x-\mu_1)} > \frac{P_2}{\sqrt{|2\pi\Sigma_2|}} e^{-\frac{1}{2}(x-\mu_2)^t \Sigma_2^{-1}(x-\mu_2)}$$
(3.76)

That inequality will generate a decision boundary that will be non-linear and will depend on the parameters of each distribution.

For some problems, a simpler decision boundary is necessary to discriminate between the classes. Normally, a linear discriminant is desired and so, one must find a decision boundary with the form of a hyperplane as

$$\mathbf{w}^t \mathbf{x} - \mathbf{b} = 0$$
(3.77)

where $\mathbf{w}$ is the coefficient of the hyperplane and $\mathbf{b}$ is the displacement of the plane. Figure 3.1 illustrates the case of two classes (represented by some samples) and a linear boundary separator. In order to obtain the optimal linear separator, Fischer's idea was to project the data on to a one-dimensional line $y = \mathbf{w}^t \mathbf{x}$ such that it maximises the distance between the projected means, keeping the variance of the projected set as small as possible. Hence, the goal is to find a projection coefficient $\mathbf{w}$ such that it maximises the difference between the mean of the

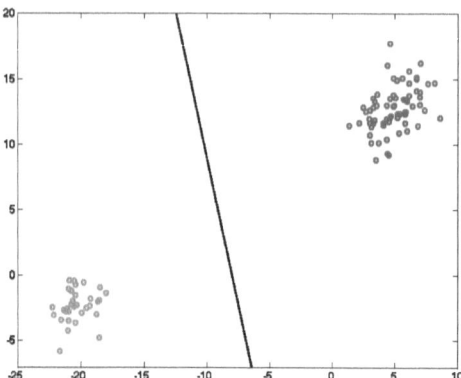

Figure 3.1: Two classes (represented by the circles) and a linear separator boundary (dark line).

projected values y_1 and y_2 (projection of the first set $\mathbf{x}_1$ and the second $\mathbf{x}_2$) and minimises the variance of the combined set $y_1; y_2$. Figure 3.2 illustrates two cases of projections where one of them is the optimal projection.

In Figure 3.2(a), it is noticeable that the projected data sets Y_1 and Y_2 have several overlapping points and that the means are not very far apart. In that case, to choose a decision boundary would inevitably produce some classification error. On the other hand, in Figure 3.2(b) the means are very well separated and practically no overlap occurs.

Notice that once projected, the data set Y (formed by samples y_1 and y_2) to be classified is one-dimensional and the decision boundary is just a threshold value chosen in the line of projection. The parameters of the distribution of the new RV Y are now (mean and variance respectively) $\mu_{i,y}$ and $\sigma^2_{i,y}$ for $i = 1, 2$

Those values can be computed from the original X_1 and X_2 data set as

$$\mu_{i,y} = \mathbf{w}^t \mu_i \tag{3.78}$$

and

$$\sigma_{i,y}^2 = \mathbf{w}^t \Sigma_i \mathbf{w} \tag{3.79}$$

with both cases for $i = 1, 2$.

That can be demonstrated using the estimators for the mean of the variable Y_i as

$$\mu_{i,y} = \frac{1}{N} \sum_{j=1}^{N} y_j$$

$$\mu_{i,y} = \frac{1}{N} \sum_{j=1}^{N} \mathbf{w}^t \mathbf{x}_j = \mathbf{w}^t \frac{1}{N} \sum_{j=1}^{N} \mathbf{x}_j = \mathbf{w}^t \mu_{i,x} \tag{3.80}$$

and for the variance as

$$\sigma^2_{i,y} = \frac{1}{N} \sum_{j=1}^{N} (y_j - \mu_{i,y})^2$$

$$\mu_{i,y} = \frac{1}{N} \sum_{j=1}^{N} (\mathbf{w}^t \mathbf{x}_j - \mathbf{w}^t \mu_i)^2 = \tag{3.81}$$

$$\mathbf{w}^t \left(\frac{1}{N} \sum_{j=1}^{N} (\mathbf{x}_j - \mu_i)^t (\mathbf{x}_j - \mu_i) \right) \mathbf{w} = \mathbf{w}^t \Sigma_i \mathbf{w}$$

Now that we have the parameters for the data set Y, we can assemble the function (that will have to be maximised)

$$J(\mathbf{w}) = \frac{(\mu_{1,y} - \mu_{2,y})^2}{\sigma_{1,y}^2 + \sigma_{2,y}^2}$$

$$J(\mathbf{w}) = \frac{(\mathbf{w}^t \mu_1 - \mathbf{w}^t \mu_2)^2}{\mathbf{w}^t \Sigma_1 \mathbf{w} + \mathbf{w}^t \Sigma_2 \mathbf{w}} \tag{3.82}$$

where the numerator represents the separation of the means and the denominator represents the overall variance of the combined data set $Y_1; Y_2$.

The goal now is to maximise $J(\mathbf{w})$ by differentiating and solving for $\mathbf{w}$. We then write (for ease of notation)

$$\mu = \mu_1 - \mu_2$$
$$\Sigma = \Sigma_1 + \Sigma_2 \tag{3.83}$$

and then proceed with the differentiation as

$$\frac{\partial J}{\partial \mathbf{w}} = \frac{2\left(\mathbf{w}^t\mu\right)\mu^t\mathbf{w}^t\Sigma\mathbf{w} - \left(\mathbf{w}^t\mu\right)^2\mathbf{w}^t\left(\Sigma^t + \Sigma\right)}{\left(\mathbf{w}^t\Sigma\mathbf{w}\right)^2} \tag{3.84}$$

Now, solving for $\mathbf{w}$, we have

$$2\left(\mathbf{w}^t\mu\right)\mu^t\left(\mathbf{w}^t\Sigma\mathbf{w}\right) - \left(\mathbf{w}^t\mu\right)^2\mathbf{w}^t\left(\Sigma^t + \Sigma\right) = 0 \tag{3.85}$$

and using the fact that the covariance is always symmetric, we can write

$$\mu^t\left(\mathbf{w}^t\Sigma\mathbf{w}\right) = \left(\mathbf{w}^t\mu\right)\mathbf{w}^t\Sigma \tag{3.86}$$

That leads to

$$\mu^t\frac{\left(\mathbf{w}^t\Sigma\mathbf{w}\right)}{\left(\mathbf{w}^t\mu\right)} = \mathbf{w}^t\Sigma \tag{3.87}$$

and can be written in short as

$$\mu\lambda = \Sigma\mathbf{w} \tag{3.88}$$

where λ is

$$\lambda = \frac{\left(\mathbf{w}^t\Sigma\mathbf{w}\right)}{\left(\mathbf{w}^t\mu\right)} \tag{3.89}$$

In principle, equation (3.87) is very hard to solve because we have a very complicated expression in $\mathbf{w}$. Nevertheless, the fact that λ is a number indicates that, whatever solution we find for $\mathbf{w}$, changing the numeric value for λ will affect only the size of the vector $\mathbf{w}$ and not its direction (that is what is important). Hence, one can ignore the constant λ (or make it a value that normalises $\mathbf{w}$) and finally write the solution as ■

$$\mathbf{w} = \lambda(\Sigma_1 + \Sigma_2)^{-1}(\mu_1 - \mu_2) \tag{3.90}$$

Equation (3.90) is the optimal projection vector that will produce new variables Y_1 and Y_2 with maximum mean separation and minimum variance.

3.5.1 Optimum Linear Discriminant

As stated at the beginning of this section, the Linear Fischer Discriminant, consists in a linear decision boundary of the form $\mathbf{w}^t \mathbf{x} - b = 0$. We then interpret this decision boundary as a one-dimensional threshold classification of a projected variable $y = \mathbf{w}^t \mathbf{x}$. We saw that the projection vector $\mathbf{w}$ can be found using (3.90). The threshold b now is a value that separates the projected values into the two classes (class 1 means $y > b$ and class 2 otherwise).

To find the value of b, we go back to assemble an hypothesis test (now in y variable) as

$$H_0 : p(y|C_1) > p(y|C_2) \tag{3.91}$$

Since we have the parameters for each class C_1 and C_2 (equation (3.78) and (3.79)) we have the expression for the probabilities $p(y|C_i)$. The decision threshold for hypothesis one will be the value where the two probabilities meet $(p(b|C_1) = p(b|C_2))$. That will produce

$$\frac{P_1}{\sqrt{2\pi}\sigma_{1,y}} e^{-\frac{(b-\mu_{1,y})^2}{2\sigma_{1,y}^2}} = \frac{P_2}{\sqrt{2\pi}\sigma_{2,y}} e^{-\frac{(b-\mu_{2,y})^2}{2\sigma_{2,y}^2}} \tag{3.92}$$

$$k_1 - \frac{(b-\mu_{1,y})^2}{2\sigma_{1,y}^2} = k_2 - \frac{(b-\mu_{2,y})^2}{2\sigma_{2,y}^2} \tag{3.93}$$

where $k_i = \log(P_i) - 0.5 \log(2\pi) - 2 \log(\sigma_{i,y})$. Developing the algebra, we finally have that the value of the threshold b is given by the solution for the equation ■

$$
\begin{aligned}
&\left(\sigma_{2,y}^2 - \sigma_{1,y}^2\right) b^2 + \\
&\left(2\sigma_{1,y}^2 \mu_{2,y} - 2\sigma_{2,y}^2 \mu_{1,y}\right) b + \\
&\sigma_{2,y}^2 \mu^2{}_{1,y} - \sigma_{1,y}^2 \mu^2{}_{2,y} - 2\left(k_1 - k_2\right) \sigma_{2,y}^2 \sigma_{1,y}^2 = 0
\end{aligned}
\tag{3.94}
$$

Notice that it is possible to have two solutions for b. That is normal, since the meeting point of two Gaussians $(p(b|C_1) = p(b|C_2))$ can occur in two different places (see Figure 3.3). As the value of b is interpreted as a threshold, the natural choice between the two solutions is to pick the one that produces the least classification error and that is the value that is between the means of the two Gaussians.

3.6   Rényi Quadratic Entropy Estimator

In recent developments (2002 to 2006), Principe et al came up with a new machine learning theory called Information Theoretic Learning (ITL). The idea is that one does not use mean square error to train a learning machine (Neural Net, RBF, etc...). Instead of the square error, what is optimised is the entropy of that error. That led to a whole new set of algorithms that are being applied successfully in all of the learning machine applications. But, for that to be possible, one needs an analytical form for the entropy of a give data set (such that one can optimise analytically). The problem with having the analytical form for the entropy of a given data set is that entropy (Shannon entropy) is given by

$$
H(X) = \int_{-\infty}^{\infty} p(x) \log(p(x)) \mathrm{d}x
\tag{3.95}
$$

Here, X represents the RV and x its numerical values. In general x is a continuous variable that goes from $-\infty$ to ∞. $p(x)$ is the probability distribution of the RV X.

In the case of using entropy for system adaptation in machine learning, the RV x is the difference (or error) between some desired set of values d_i and the values produced by the machine that is being adapted $y_i(\mathbf{w})$ ($e_i = d_i - y_i(\mathbf{w})$). The vector $\mathbf{w}$ is the set of parameters that will be adapted to "train" the machine. In this scenario, some problems arise when one wants to adapt $\mathbf{w}$ using the entropy (that can now be written as $H(\mathbf{w})$, since now X is the RV with samples given by e_i and this is a function of $\mathbf{w}$). Basically, there are two problems. The first problem is that one does not have the expression for $p(e)$ (probability of error). Instead what we have are some samples from its distribution (e_i, $i = 0, 1, 2, \ldots, N-1$). Because of that, the entropy cannot be stated as a function of $\mathbf{w}$ to be optimised (for instance differentiated with respect to $\mathbf{w}$). This first difficulty is overcome by using an estimation for $p(e)$ from its samples. Here we use Parzen's estimator, given by

$$p(x) = \frac{1}{N} \sum_{i=0}^{N-1} k_\sigma(x - x_i) \tag{3.96}$$

where $k_\sigma(.)$ is a kernel function that must meet some requirements[xx][1]. The subindex σ is the "length" of the kernel and is a given parameter.

Now, it is (in principle) possible to have an expression for the entropy using (3.96) in the expression for the entropy (3.95). But that leads to the second main problem of using entropy as an optimisation criterion. The integral in (3.95) using (3.96) hase no simple solution. To overcome this new difficulty, one must turn to other definitions for entropy. One definition that can be used is Renyi's entropy, given by

---

[1] In general, $k_\sigma(.)$ must have the form of a zero mean pdf and σ is related with its second moment.

$$H_r(X) = \frac{1}{1-\alpha} \log \int_{-\infty}^{\infty} p(x)^\alpha dx \qquad (3.97)$$

where α is a parameter that changes the scale of the entropy[2]. In order to use Renyi's entropy with the Parzen estimation for $p(x)$, we will use the case for $\alpha = 2$ and then compute the quadratic Renyi's entropy as

$$H_r(X) = -\log \int_{-\infty}^{\infty} p(x)^2 dx \qquad (3.98)$$

Now, it is possible to plug the estimation (3.96) into the entropy expression (3.98) and have

$$H_r(X) = -\log \int_{-\infty}^{\infty} \left(\frac{1}{N} \sum_{i=0}^{N-1} k_\sigma(x - x_i) \right)^2 dx \qquad (3.99)$$

Now, the integral can be solved by rewriting the square of the summation as a product of summations with two indices as

$$H_r(X) = -\log \frac{1}{N^2} \int_{-\infty}^{\infty} \sum_{i=0}^{N-1} k_\sigma(x - x_i) \sum_{j=0}^{N-1} k_\sigma(x - x_j) dx$$

$$(3.100)$$

Now, combining into a double summation and bringing the integral inside the sum leads to

$$H_r(X) = -\log \frac{1}{N^2} \sum_{i=0}^{N-1} \int_{-\infty}^{\infty} k_\sigma(x - x_j)k_\sigma(x - x_i) dx \qquad (3.101)$$

---

[2]In the next section it is shown that for $\alpha = 1$ the Renyi entropy assumes the same form as Shannon entropy, therefore being a general measure of entropy.

One of the common kernels used in the Parzen estimates is the Gaussian kernel. Plugging the expression for the Gaussian into the entropy expression, we have

$$H_r(X) = -\log \frac{1}{N^2} \sum_{i=0}^{N-1} \int_{-\infty}^{\infty} \frac{1}{\sqrt{2\pi}\sigma} e^{-\frac{1}{2\sigma^2}(x-x_i)^2} \frac{1}{\sqrt{2\pi}\sigma} e^{-\frac{1}{2\sigma^2}(x-x_j)^2} dx$$

(3.102)

Which, carrying on the algebra, leads to

$$H_r(X) = -\log \frac{1}{N^2} \sum_{i=0}^{N-1} \frac{1}{2\pi\sigma^2} \int_{-\infty}^{\infty} e^{-\frac{1}{2\sigma^2}\left((x-x_i)^2+(x-x_j)^2\right)} dx$$

(3.103)

Now, the integral can be solved (see section 3.1) and we get

$$H_r(X) = -\log \frac{1}{N^2} \sum_{i=0}^{N-1} \frac{1}{2\pi\sigma^2} \sqrt{\pi}\sigma e^{-\frac{1}{4\sigma^2}(x_i-x_j)^2}$$

(3.104)

$$H_r(X) = -\log \frac{1}{N^2} \sum_{i=0}^{N-1} \frac{1}{2\sqrt{\pi}\sigma} e^{-\frac{1}{4\sigma^2}(x_i-x_j)^2}$$

It is possible to rewrite σ such that we have

$$H_r(X) = -\log \frac{1}{N^2} \sum_{i=0}^{N-1} \frac{1}{\sqrt{2\pi}\sqrt{2}\sigma} e^{-\frac{1}{2(\sqrt{2}\sigma)^2}(x_i-x_j)^2}$$

(3.105)

and write it in a simpler form as ■

$$H_r(X) = -\log \frac{1}{N^2} \sum_{i=0}^{N-1} k_{\sqrt{2}\sigma}(x_i - x_j)$$

(3.106)

which is an estimator for entropy using only the sample points from a RV X. What is remarkable in this result is that now there exists an analytical expression for the entropy of a RV when one has only some sample points from that RV. Since (for the machine learning paradigm) the samples will be the error between some desired known value and a parameter **w**-dependent output, one can optimise the entropy with respect to the parameters. That, as said before, gave rise to a whole new paradigm to solve machine learning problems.

3.7 Renyi entropy for $\alpha = 1$

In this section we will show that Shannon entropy is a particular case of *Renyi's* entropy where the parameter $\alpha = 1$. First, it is important to see that the case of $\alpha = 1$ is a limiting case. If one just substitutes the value 1 into (3.97), we have $\log(1)/0$, which is an indetermination. Because of that fact, one must make the limit

$$H_1(X) = \lim_{\alpha \to 1} \frac{\log \int\limits_{-\infty}^{\infty} p(x)^{\alpha} \mathrm{d}x}{1 - \alpha} \qquad (3.107)$$

Now, using L'Hôpital's Rule and differentiating the numerator and denominator, we have

$$H_1(X) = \frac{\lim_{\alpha \to 1} \dfrac{\int\limits_{-\infty}^{\infty} \frac{\partial p(x)^{\alpha}}{\partial \alpha} \mathrm{d}x}{\int\limits_{-\infty}^{\infty} p(x)^{\alpha} \mathrm{d}x}}{-1} \qquad (3.108)$$

Proceeding with the differentiation, we have

$$H_1(X) = - \int\limits_{-\infty}^{\infty} p(x) \log(p(x)) \mathrm{d}x \qquad (3.109)$$

which is the definition in (3.95) ■.

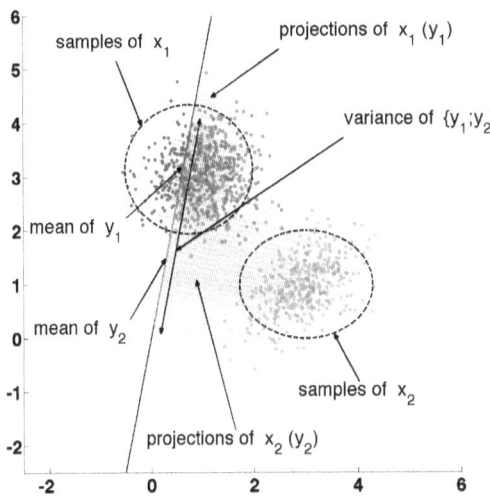

(a) Projection into a non-optimal line

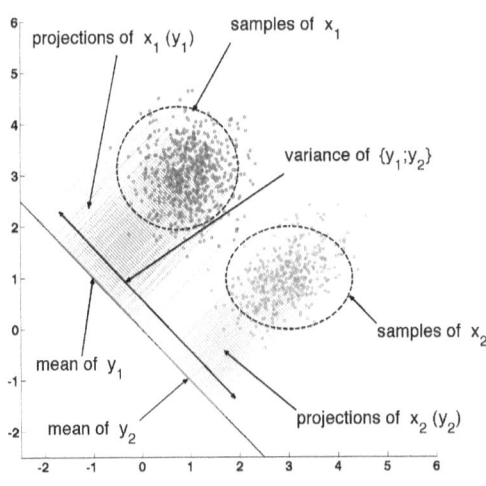

(b) Projection into the optimal line

Figure 3.2: Projections on two different lines.

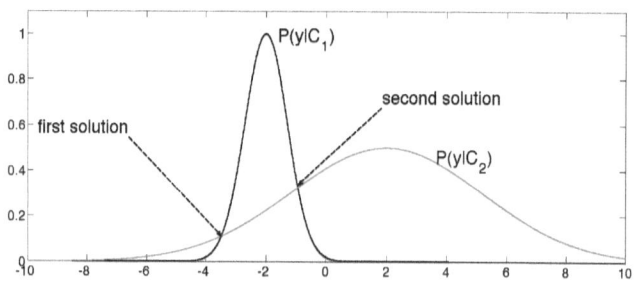

Figure 3.3: Two gaussians with two equal value points.

Chapter 4

Signal Processing

Generally, engineering students working in areas such as signal processing need to deal with some mathematics and make use of many formulae. In this chapter, we will try to show the derivation for several signal-processing-related formulae and equations. As in the rest of the book, we will cover only a small portion of the formulae and expressions used. But, again, these are a selection that might help the reader in some application.

4.1 Fourier transforms

The Fourier transform is normally presented in books as separate transforms depending on the nature of the signal you are using. If the signal is continuous and periodic, we have Fourier series; if it is discrete and aperiodic, we have the discrete Fourier series; if the signal is discrete and periodic, we have the discrete Fourier transform and if we have a continuous and aperiodic signal, we have the original Fourier transform.

In the following sections we are going to derive all the forms of the Fourier transform (series, discrete series and discrete transform), starting with the definition of the continuous transform. It is actually a more formal and natural way to look at the Fourier transform because it does not involve many formulae.

Instead, we will show that all the forms of Fourier transforms that are presented in the books are indeed the definition rewritten to use the signal characteristics (periodicity and continuity). In fact, we will even derive the inverse from the definition.

4.1.1   Fourier transform

Starting from the definition, the Fourier transform will be stated as (this will be the only definition used, all the other forms will derive from this)

$$X(f) \triangleq \int_{-\infty}^{\infty} x(t)e^{-j2\pi ft}\mathrm{d}t \qquad (4.1)$$

Its inverse can be obtained by the following procedure: Starting from the definition, we do

$$X(f) = \int_{-\infty}^{\infty} x(t)e^{-j2\pi ft}\mathrm{d}t$$

$$e^{j2\pi ft'}X(f) = \int_{-\infty}^{\infty} x(t)e^{-j2\pi ft}e^{j2\pi ft'}\mathrm{d}t$$

$$\int_{-\infty}^{\infty} e^{j2\pi ft'}X(f)\mathrm{d}f = \int_{-\infty}^{\infty}\int_{-\infty}^{\infty} x(t)e^{-j2\pi ft}e^{j2\pi ft'}\mathrm{d}t\mathrm{d}f$$

$$\int_{-\infty}^{\infty} e^{j2\pi ft'}X(f)\mathrm{d}f = \int_{-\infty}^{\infty} x(t)\int_{-\infty}^{\infty} e^{-j2\pi f(t-t')}\mathrm{d}f\mathrm{d}t$$

$$(4.2)$$

It is possible to show that[1]

---

[1]There are several ways to prove this expression. One version will be shown in the next sections.

$$\int_{-\infty}^{\infty} e^{-j2\pi f(t-t')} \mathrm{d}f = \delta(t - t') \tag{4.3}$$

That leads to

$$\int_{-\infty}^{\infty} e^{j2\pi f t'} X(f)\mathrm{d}f = \int_{-\infty}^{\infty} x(t)\delta(t - t')\mathrm{d}t = x(t') \tag{4.4}$$

and we conclude that[2]

$$x(t) = \int_{-\infty}^{\infty} X(f)e^{j2\pi f t}\mathrm{d}f \tag{4.5}$$

which is the form of the inverse Fourier formula.

4.1.2   Discrete Fourier series

If the signal is discrete (discretised with sampling period T) and aperiodic in time, we can represent that signal as

$$x(t) = \sum_{n=-\infty}^{\infty} x[n]\delta(t - nT) \tag{4.6}$$

Now, using the definition to compute the Fourier transform of that signal, we have

---

[2]Notice that in the initial definition, the inverse is derived without normalisation. Some definitions utilise the value of the transform normalised to 2π. In this case we must pay attention to use the variable ω (angular frequency). Besides, one must lead to another having only one definition.

$$X(f) = \int_{-\infty}^{\infty} \left(\sum_{n=-\infty}^{\infty} x[n]\delta(t - nT) \right) e^{-j2\pi f t} dt$$

$$= \sum_{n=-\infty}^{\infty} x[n] \int_{-\infty}^{\infty} \delta(t - nT) e^{-j2\pi f t} dt \qquad (4.7)$$

$$X(f) = \sum_{n=-\infty}^{\infty} x[n] e^{-j2\pi f n T}$$

That is a continuous and periodic signal (with period $\frac{1}{T}$). We can represent the inverse Fourier transform (the signal itself) by $x[n]$ and we can compute it in the following way.

$$X(f) = \sum_{n=-\infty}^{\infty} x[n] e^{-j2\pi f n T}$$

$$X(f) e^{j2\pi f n'T} = \sum_{n=-\infty}^{\infty} x[n] e^{-j2\pi f n T} e^{j2\pi f n'T}$$

$$\int_{0}^{\frac{1}{T}} X(f) e^{j2\pi f n'T} df = \sum_{n=-\infty}^{\infty} x[n] \int_{0}^{\frac{1}{T}} e^{-j2\pi f (n-n')T} df \qquad (4.8)$$

$$\int_{0}^{\frac{1}{T}} X(f) e^{j2\pi f n'T} df = \sum_{n=-\infty}^{\infty} x[n] \left(\frac{e^{-j2\pi f (n-n')T}}{-j2\pi (n - n')T} \right) \Bigg|_{0}^{\frac{1}{T}}$$

$$\int_{0}^{\frac{1}{T}} X(f) e^{j2\pi f n'T} df = \sum_{n=-\infty}^{\infty} x[n] \frac{e^{-j2\pi (n-n')} - 1}{-j2\pi (n - n')T}$$

Analysing the term

$$\sum_{n=-\infty}^{\infty} x[n] \frac{e^{-j2\pi (n-n')} - 1}{-j2\pi (n - n')T} \qquad (4.9)$$

we notice that[3]

$$\frac{e^{-j2\pi(n-n')} - 1}{-j2\pi(n - n')T} = \begin{cases} \frac{1}{T}, & n = n' \\ 0, & n \neq n' \end{cases} \qquad (4.10)$$

and therefore the summation becomes

$$\sum_{n=-\infty}^{\infty} x[n] \frac{e^{-j2\pi(n-n')} - 1}{-j2\pi(n - n')T} = \frac{1}{T}x[n'] \qquad (4.11)$$

That leads us to the final inverse formula

$$x[n] = T \int_{0}^{\frac{1}{T}} X(f)e^{j2\pi f nT} \mathrm{d}f \qquad (4.12)$$

4.1.3 Fourier series

If the signal is discrete in the frequency domain (discretised in frequency with period f_0)[4] and aperiodic, we can write its Fourier transform as

$$X(f) = \sum_{k=-\infty}^{\infty} X[k]\delta(f - kf_0) \qquad (4.13)$$

,

and applying the inverse Fourier transform we obtain

---

[3]Using L'Hôpital's rule.

[4]Notice that the idea of "period" in the frequency domain requires attention. A function that is periodic (or samples periodically) in the frequency domain repeats itself at each multiple of a frequency (since the x axis has frequency units) and, therefore, the period of a periodic frequency function in the frequency domain is itself a value of frequency (not time).

$$x(t) = \int_{-\infty}^{\infty} \left(\sum_{k=-\infty}^{\infty} X[k]\delta(f - kf_0) \right) e^{j2\pi ft} df$$

$$= \sum_{k=-\infty}^{\infty} X[k] \int_{-\infty}^{\infty} \delta(f - kf_0) e^{j2\pi ft} df \qquad (4.14)$$

$$x(t) = \sum_{k=-\infty}^{\infty} X[k] e^{jk2\pi f_0 t}$$

That is continuous and periodic (with period equal to $\frac{1}{f_0}$). Its Fourier transform can be represented simply by $X[k]$ and can be computed as

$$x(t) = \sum_{k=-\infty}^{\infty} X[k] e^{j2\pi f_0 k t}$$

$$x(t)e^{-j2\pi f_0 k't} = \sum_{k=-\infty}^{\infty} X[k] e^{j2\pi f_0 k t} e^{-j2\pi f_0 k't}$$

$$\int_0^{\frac{1}{f_0}} x(t)e^{-j2\pi f_0 k't} dt = \int_0^{\frac{1}{f_0}} \sum_{k=-\infty}^{\infty} X[k] e^{j2\pi f_0 k t} e^{-j2\pi f_0 k't} dt$$

$$\int_0^{\frac{1}{f_0}} x(t)e^{-j2\pi f_0 k't} dt = \sum_{k=-\infty}^{\infty} X[k] \int_0^{\frac{1}{f_0}} e^{-j2\pi f_0 (k'-k)t} dt \qquad (4.15)$$

$$\int_0^{\frac{1}{f_0}} x(t)e^{-j2\pi f_0 k't} dt = \sum_{k=-\infty}^{\infty} X[k] \left. \frac{e^{-j2\pi f_0 (k'-k)t}}{-j2\pi f_0 (k'-k)} \right|_0^{\frac{1}{f_0}}$$

$$\int_0^{\frac{1}{f_0}} x(t)e^{-j2\pi f_0 k't} dt = \sum_{k=-\infty}^{\infty} X[k] \frac{e^{-j(k'-k)2\pi} - 1}{-j2\pi f_0 (k'-k)}$$

With the same procedure used in the discrete Fourier series for the summation, we notice that the inverse is given by

$$X[k] = f_0 \int_0^{\frac{1}{f_0}} x(t)e^{-j2\pi f_0 k t} dt \tag{4.16}$$

4.2 Discrete Fourier transform

Finally, if the signal is discrete and periodic in time (with period T_0), we have the following situation. Initially, suppose a signal that is limited in time (discretised with sampling period T and length $T_0 = NT$, where N is the number of samples and the rest of the signal is zero)

$$\hat{x}(t) = \sum_{n=0}^{N-1} x[n]\delta(t - nT) \tag{4.17}$$

Now, the periodic signal can be formed using $\hat{x}(t)$ as

$$x(t) = \sum_{m=-\infty}^{\infty} \hat{x}(t - T_0 m) \tag{4.18}$$

,

That leads us to a generic representation of a discrete and periodic signal as

$$x(t) = \sum_{m=-\infty}^{\infty} \sum_{n=0}^{N-1} x[n]\delta(t - T_0 m - nT) \tag{4.19}$$

Its Fourier transform can be computed as

$$X(f) = \int_{-\infty}^{\infty} \left(\sum_{m=-\infty}^{\infty} \sum_{n=0}^{N-1} x[n]\delta(t - T_0 m - nT) \right) e^{-j2\pi f t} dt$$

$$= \sum_{m=-\infty}^{\infty} \sum_{n=0}^{N-1} x[n] \int_{-\infty}^{\infty} \delta(t - T_0 m - nT) e^{-j2\pi f t} dt$$

$$= \sum_{m=-\infty}^{\infty} \sum_{n=0}^{N-1} x[n] e^{-j2\pi f (T_0 m + nT)}$$

$$= \sum_{m=-\infty}^{\infty} \sum_{n=0}^{N-1} x[n] e^{-j2\pi f T_0 m} e^{-j2\pi f T n}$$

$$= \sum_{n=0}^{N-1} x[n] e^{-j2\pi f T n} \sum_{m=-\infty}^{\infty} e^{-j2\pi f T_0 m}$$

$$(4.20)$$

Using the property of the function "Dirac comb", we can identify[5],[6]

$$\frac{1}{T_0} \sum_{m=-\infty}^{\infty} \delta\left(f - \frac{m}{T_0} \right) = \sum_{m=-\infty}^{\infty} e^{-j2\pi f T_0 m} \qquad (4.21)$$

and obtain

$$X(f) = \sum_{n=0}^{N-1} x[n] e^{-j2\pi f T n} \frac{1}{T_0} \sum_{m=-\infty}^{\infty} \delta\left(f - \frac{m}{T_0} \right)$$

$$= \frac{1}{T_0} \sum_{m=-\infty}^{\infty} \sum_{n=0}^{N-1} x[n] e^{-j\frac{2\pi m n T}{T_0}} \delta\left(f - \frac{m}{T_0} \right)$$

$$(4.22)$$

---

[5]This expression. Can be proved writing the Fourier series of the Dirac comb and verifying that all coefficients are equal to $\frac{1}{T}$.

[6]This property comes from the Fourier transform of the "Dirac comb" that is defined as the sum of Dirac deltas spaced in time. Its Fourier transform also results in a Dirac comb spaced in frequency.

That can be represented by the discrete frequencies $\frac{m}{T_0}$ only, as (using k as discrete frequency instead of m)

$$X[k] = \frac{1}{T_0} \sum_{n=0}^{N-1} x[n] e^{-j\frac{2\pi n kT}{T_0}} \tag{4.23}$$

Since $T_0 = NT$, we have

$$X[k] = \frac{1}{NT} \sum_{n=0}^{N-1} x[n] e^{-j\frac{2\pi n k}{N}} \tag{4.24}$$

It is very common to consider the sampling time equal to unity[7], and that leads us to the expression for the discrete Fourier transform as

$$X[k] = \frac{1}{N} \sum_{n=0}^{N-1} x[n] e^{-j\frac{2\pi n k}{N}} \tag{4.25}$$

That is also a periodic discrete signal (with period also equal to N)

The inverse can be obtained as follows (still considering the unitary sampling time)

$$X[k] = \frac{1}{N} \sum_{n=0}^{N-1} x[n] e^{-j\frac{2\pi n k}{N}}$$

$$e^{j\frac{2\pi k n'}{N}} X[k] = \frac{1}{N} \sum_{n=0}^{N-1} x[n] e^{-j\frac{2\pi n k}{N}} e^{j\frac{2\pi n' k}{N}}$$

$$\sum_{k=0}^{N-1} e^{j\frac{2\pi k n'}{N}} X[k] = \frac{1}{N} \sum_{k=0}^{N-1} \sum_{n=0}^{N-1} x[n] e^{-j\frac{2\pi n k}{N}} e^{j\frac{2\pi n' k}{N}}$$

$$\sum_{k=0}^{N-1} e^{j\frac{2\pi k n'}{N}} X[k] = \frac{1}{N} \sum_{n=0}^{N-1} x[n] \sum_{k=0}^{N-1} e^{-j\frac{2\pi (n-n') k}{N}} \tag{4.26}$$

---

[7]This consideration is acceptable when we work with both transforms, direct and inverse, in their discrete form. In this manner, either time or frequency is represented by integer indices that are dimensionless.

Analysing the expression $\sum_{k=0}^{N-1} e^{-j\frac{2\pi(n-n')k}{N}}$, we realise that, if $n \neq n'$, we have

$$\sum_{k=0}^{N-1} r^k = \frac{r^N - 1}{r - 1}$$

$$r = e^{-j\frac{2\pi(n-n')}{N}}$$

$$\sum_{k=0}^{N-1} e^{-j\frac{2\pi(n-n')k}{N}} = \frac{e^{-j\frac{2\pi(n-n')N}{N}} - 1}{e^{-j\frac{2\pi(n-n')}{N}} - 1} = \frac{e^{-j2\pi(n-n')} - 1}{e^{-j\frac{2\pi(n-n')}{N}} - 1} = 0$$

$$(4.27)$$

and when $n = n'$, we have

$$\sum_{k=0}^{N-1} e^{-j\frac{2\pi(n-n)k}{N}} = \sum_{k=0}^{N-1} 1 = N \qquad (4.28)$$

Therefore, the inverse discrete Fourier transform becomes

$$x[n] = \sum_{k=0}^{N-1} X[k] e^{j\frac{2\pi nk}{N}} \qquad (4.29)$$

4.3   Wiener Filter denoising

One of the first adaptive filters that came out after the popularisation of computers was the Wiener Filter. The term "adaptive" means that the filter is adjusted to perform the filtering by adapting the filter coefficients (weights) using some knowledge of the signal one wishes to filter. One of the uses for adaptive filtering is to filter out the noise in a signal. In this section we will derive the formula for the Wiener Filter when used to perform denoising.

Let $s[n]$ and $\eta[n]$ be a signal and a zero mean Gaussian noise with variance σ_η^2. We will use the signal $x[n]$ given by (through the paper we will refer to the samples of a signal as the signal itself, for simplicity's sake)

$$x[n] = s[n] + \eta[n] \tag{4.30}$$

and the variance σ_η^2 of the noise as the only information available for denoising. Let w_i be the coefficients of some linear filter and $y[n]$ its output given by the following convolution

$$y[n] = \sum_i w_i x[n-i] \tag{4.31}$$

In the denoising problem, we wish to find a set of w_i which makes the output $y[n]$ of the filter equal to the signal $s[n]$ (notice that we donâĂŹt have access to $s[n]$). In fact, we should have

$$s[n] = \sum_i w_i x[n-i] \tag{4.32}$$

Let us multiply both sides of the equation by $x[n-k]$ and take the expected value. We get

$$E\left\{x[n-k]s[n]\right\} = \sum_i w_i E\left\{x[n-i]x[n-k]\right\} \tag{4.33}$$

The autocorrelation function of $x[n]$ is what appears on the right-hand side. Now, since we do not have access to $s[n]$, let us use equation (4.30) in (4.33) and proceed as

$$E\left\{x[n-k]\left(x[n] - \eta[n]\right)\right\} = \sum_i w_i r[i-k]$$

$$E\left\{x[n-k]x[n]\right\} - E\left\{x[n-k]\eta[n]\right\} = \sum_i w_i r[i-k] \tag{4.34}$$

where $r[i]$ is the autocorrelation function of the variable $x[n]$. Now using equation (4.30) again for $x[n]$ and the fact that the noise is uncorrelated, we get

$$r[k] - \sigma_\eta^2 \delta[k] = \sum_i w_i r[i-k] \tag{4.35}$$

where $\delta[n]$ is one for $n = 0$ and zero otherwise. Equation (4.35) must be valid for all k, leading to a system of linear equations. Solving this system, leads to the Wiener solution for the denoising problem. Notice that no signal $s[n]$ is needed and perfect denoising needs a large number of samples. The only extra information needed is the noise variance σ_η^2. Depending on the method used to solve the system, we get the well-known methods such as Least Mean Squares, Recursive Least Squares etc [1].

4.4   Kalman Filter

The challenge with the Kalman Filter is to estimate the state $\mathbf{x}$ of a system given the previous state and the new measured output. This problem is well solved by linear estimators like the Luenberger observer, among others. The problem arises when the measures (or the states) are contaminated with noise. The Kalman filter estimates the new states so as to minimise the covariance or energy of the error in the state. Doing so, it is optimal in the statistical sense.

Let us consider the following system

$$\begin{aligned} \mathbf{x}_k &= \mathbf{F}\mathbf{x}_{k-1} + \mathbf{w}_{k-1} \\ \mathbf{y}_k &= \mathbf{H}^t\mathbf{x}_k + \mathbf{v}_k \end{aligned} \qquad (4.36)$$

where $\mathbf{F}$ is the transition matrix, $\mathbf{x}_k$ and $\mathbf{y}_k$ are the state vector and output at instant k. $\mathbf{w}_{k-1}$ and $\mathbf{v}_{k-1}$ are the state noise and output noise and $\mathbf{H}$ is the measurement vector (or matrix). The noise is assumed to be zero mean. We call that the measured system or real value system. Those are the variables that are actually flowing in the system. From those, we know the transition and measurement matrix and also we can gain access to the measured output. The samples of the noise are not known, but we do know their covariance matrices $\mathbf{W}$ and $\mathbf{V}$. The problem now is to estimate the state vector given the previous estimate and the measured output. Let $\mathbf{x}_{k|k-1}$ and

$\mathbf{y}_{k|k-1}$ be the estimatives of the value of the state and output given the previous estimative of the states. We can now estimate the states and output as

$$\mathbf{x}_{k|k-1} = \mathbf{F}\mathbf{x}_{k-1|k-1}$$
$$\mathbf{y}_{k|k-1} = \mathbf{H}^t\mathbf{x}_{k|k-1} \tag{4.37}$$

Notice that as an estimative, we donâĂŹt include the noise samples and, since the noise mean is zero, we just estimate without the noise.

The Kalman approach is to correct the estimative of the state (based on the current estimative) as the current state plus an "innovation" factor. We then have

$$\mathbf{x}_{k|k} = \mathbf{x}_{k|k-1} + \mathbf{G}(\mathbf{y}_k - \mathbf{y}_{k|k-1}) \tag{4.38}$$

The inclusion of the concept of innovation is due to the fact that the second term in the correction of the estimative is a gain times the difference (thus call innovation) in the output. This is so that if the current output matches the estimative, no correction must be done in the estimated state.

The problem now is to find $\mathbf{G}$ such that this correction produces the smaller square error state estimate $\|\mathbf{x}_k - \mathbf{x}_{k|k}\|^2$. To do that, we must define the covariance

$$\mathbf{P}_{k|k-1} = \text{cov}\{\mathbf{x}_k - \mathbf{x}_{k|k-1}\} \tag{4.39}$$

and

$$\mathbf{P}_{k|k} = \text{cov}\{\mathbf{x}_k - \mathbf{x}_{k|k}\} \tag{4.40}$$

where the first expression is the covariance of the estimative and the second is the covariance of the correction. The goal is to minimise the trace of the covariance of the correction error (thus minimising the square norm of the error) and compute that using the covariance of the estimate.

To compute $\mathbf{P}_{k|k-1}$ we substitute the expressions of the measured state and estimated state as

$$\begin{aligned} \mathbf{P}_{k|k-1} &= \operatorname{cov}\{\mathbf{F}\mathbf{x}_{k-1} + \mathbf{w}_{k-1} - \mathbf{F}\mathbf{x}_{k-1|k-1}\} = \\ &\quad \operatorname{cov}\{\mathbf{F}(\mathbf{x}_{k-1} - \mathbf{x}_{k-1|k-1}) + \mathbf{w}_{k-1}\} \end{aligned} \tag{4.41}$$

Using the fact that the noise is uncorrelated to the state and state estimates, we have

$$\mathbf{P}_{k|k-1} = \mathbf{F}\operatorname{cov}\{\mathbf{x}_{k-1} - \mathbf{x}_{k-1|k-1}\}\mathbf{F}^t + \mathbf{W} \tag{4.42}$$

that is equal to

$$\mathbf{P}_{k|k-1} = \mathbf{F}\mathbf{P}_{k-1|k-1}\mathbf{F}^t + \mathbf{W} \tag{4.43}$$

Now, we compute the covariance of the correction $\mathbf{P}_{k|k}$ by substituting the correction term in the expression for the covariance. We now have

$$\mathbf{P}_{k|k} = \operatorname{cov}\{\mathbf{x}_k - \mathbf{x}_{k|k-1} - \mathbf{G}(\mathbf{y}_k - \mathbf{y}_{k|k-1})\} \tag{4.44}$$

Substituting the expression for the output and estimate, we have

$$\mathbf{P}_{k|k} = \operatorname{cov}\{\mathbf{x}_k - \mathbf{x}_{k|k-1} - \mathbf{G}(\mathbf{H}^t\mathbf{x}_k + \mathbf{v}_k - \mathbf{H}^t x_{k|k-1})\} \tag{4.45}$$

Rewriting in terms of the state errors, we have

$$\mathbf{P}_{k|k} = \operatorname{cov}\{(\mathbf{I} - \mathbf{G}\mathbf{H}^t)(\mathbf{x}_k - \mathbf{x}_{k|k-1}) + \mathbf{G}\mathbf{v}_k\} \tag{4.46}$$

and since the output error is also uncorrelated with the states and estimates, we have

$$\mathbf{P}_{k|k} = (\mathbf{I} - \mathbf{G}\mathbf{H}^t)\mathbf{P}_{k|k-1}(\mathbf{I} - \mathbf{G}\mathbf{H}^t)^t + \mathbf{G}\mathbf{V}\mathbf{G}^t \tag{4.47}$$

Now, we must choose the gain $\mathbf{G}$ such that the trace of $\mathbf{P}_{k|k}$ is minimised. That is equivalent to differentiating $\mathbf{P}_{k|k}$ with respect to $\mathbf{G}$ and setting to zero. So we have (with $\mathbf{P}_{k|k-1}$ and $\mathbf{G}\,\mathbf{H}^t$ being symmetrical)

$$\mathbf{P}_{k|k} = \mathbf{P}_{k|k-1} - \mathbf{G}\mathbf{H}^t\mathbf{P}_{k|k-1} - \mathbf{P}_{k|k-1}\mathbf{H}\mathbf{G} + \mathbf{G}\mathbf{H}^t\mathbf{P}_{k|k-1}\mathbf{H}\mathbf{G}^t + \mathbf{G}\mathbf{V}\mathbf{G}^t$$
$$\mathbf{P}_{k|k} = \mathbf{P}_{k|k-1} - 2\mathbf{G}\mathbf{H}^t\mathbf{P}_{k|k-1} + \mathbf{G}(\mathbf{H}^t\mathbf{P}_{k|k-1}\mathbf{H} + \mathbf{V})\mathbf{G}^t$$

$$(4.48)$$

Differentiating with respect to $\mathbf{G}$ and setting to zero, we have

$$0 = -2\mathbf{H}^t\mathbf{P}_{k|k-1} + \mathbf{G}(\mathbf{H}^t\mathbf{P}_{k|k-1}\mathbf{H} + \mathbf{V}) \qquad (4.49)$$

So, the gain is computed as

$$\mathbf{G} = \mathbf{H}^t\mathbf{P}_{k|k-1}(\mathbf{H}^t\mathbf{P}_{k|k-1}\mathbf{H} + \mathbf{V})^{-1} \qquad (4.50)$$

and is well known as Kalman Gain.

Although the Kalman Gain is an expression that involves inversion of matrices, it can also be computed recursively using the inversion lemma, as the matrix to be inverted is in the form $\mathbf{u}^t\mathbf{A}\mathbf{u} + \mathbf{M}$.

4.5 Matched filtering

In digital communications one useful type of filter is the matched filter. In this filter, the input signal is supposed to be formed by a signal component $s[k]$ and a noise component $n[k]$. Hence we have the following convolution for the filter output

$$y[k] = \sum_i h[i-k]x[i] \qquad (4.51)$$

where

$$x[k] = s[k] + n[k] \qquad (4.52)$$

The goal is to find the "best" impulse response $h[k]$ for the filter. To derive that solution, we will use a more convenient notation and write the current output as (dropping the time variable)

$$y = \mathbf{h}^t\mathbf{s} + \mathbf{h}^t\mathbf{n} \tag{4.53}$$

Representing the signal part y_s and the noise part y_n

$$y = y_s + y_n \tag{4.54}$$

The "best" $\mathbf{h}$ will be the one that maximises the signal-to-noise ratio (SNR), defined as [8]

$$
\begin{aligned}
SNR &= \frac{E\left\{y_s^2\right\}}{E\left\{y_n^2\right\}} \\
&= \frac{\left(\mathbf{h}^t\mathbf{s}\right)^2}{E\left\{\left(\mathbf{h}^t\mathbf{n}\right)^2\right\}}
\end{aligned} \tag{4.55}
$$

Using the definition for an autocorrelation matrix and rewriting the inner product squares, we have

$$
\begin{aligned}
SNR &= \frac{\left(\mathbf{h}^t\mathbf{s}\right)^2}{E\left\{\mathbf{h}^t\mathbf{n}\mathbf{n}^t\mathbf{h}\right\}} \\
&= \frac{\left(\mathbf{h}^t\mathbf{s}\right)^2}{\mathbf{h}^t E\left\{\mathbf{n}\mathbf{n}^t\right\}\mathbf{h}} \\
&= \frac{\mathbf{h}^t\mathbf{s}\mathbf{s}^t\mathbf{h}}{\mathbf{h}^t\mathbf{R}_n\mathbf{h}}
\end{aligned} \tag{4.56}
$$

Now, we have to have some normalisation in the SNR in order to avoid trivial solutions ($\mathbf{h} = \mathbf{0}$). We could normalise the output noise power by doing

$$\mathbf{h}^t\mathbf{R}_n\mathbf{h} = 1 \tag{4.57}$$

and then define the function based on the constrained maximisation problem

---

[8]Notice that the numerator is deterministic and we can drop the expected value operator.

$$J = \mathbf{h}^t \mathbf{s}\mathbf{s}^t \mathbf{h} + c(\mathbf{h}^t \mathbf{R}_n \mathbf{h} - 1) \tag{4.58}$$

Now, we do

$$\frac{\partial J}{\partial \mathbf{h}} = 0 \tag{4.59}$$

and get

$$\mathbf{h}^t \mathbf{s}\mathbf{s}^t + c\mathbf{h}^t \mathbf{R}_n = 0 \tag{4.60}$$

which results in the following generalised eigenvalue problem

$$\mathbf{s}\mathbf{s}^t \mathbf{h} = \lambda \mathbf{R}_n \mathbf{h} \tag{4.61}$$

To find $\mathbf{h}$ we rewrite 4.61 as

$$\mathbf{R}_n^{-1} \mathbf{s}\mathbf{s}^t \mathbf{h} = \lambda \mathbf{R}_n^{-1} \mathbf{R}_n \mathbf{h} \tag{4.62}$$

and get

$$\mathbf{s}\mathbf{s}^t \mathbf{R}_n^{-1} \mathbf{h} = \lambda \mathbf{h} \tag{4.63}$$

By inspection of 4.61 and 4.63 we have a candidate solution as

$$\mathbf{h} = \alpha \mathbf{R}_n^{-1} \mathbf{s} \tag{4.64}$$

To verify that this is a solution, we substitute 4.64 into 4.61 as

$$\mathbf{s}\mathbf{s}^t \left(\mathbf{R}_n^{-1} \mathbf{s} \right) = \lambda \mathbf{R}_n \left(\mathbf{R}_n^{-1} \mathbf{s} \right) \tag{4.65}$$

and verify that indeed we get equation 4.63 back. Now we must find the value of α. We can multiply 4.64 by $\mathbf{h}^t$ on both sides and get

$$\mathbf{h}^t \mathbf{R}_n \mathbf{h} = \alpha \mathbf{h}^t \mathbf{s} \tag{4.66}$$

We identify the left-hand side as the normalisation definition and get

$$1 = \alpha \mathbf{h}^t \mathbf{s}$$

$$\alpha = \frac{1}{\alpha \mathbf{h}^t \mathbf{s}} \tag{4.67}$$

Substituting 4.64 we have

$$\alpha = \frac{1}{\alpha \mathbf{s}^t \mathbf{R}_n^{-1} \mathbf{s}}$$

$$\alpha = \frac{1}{\sqrt{\mathbf{s}^t \mathbf{R}_n^{-1} \mathbf{s}}} \tag{4.68}$$

which makes the solution for the matched filter be ■

$$\mathbf{h} = \frac{1}{\sqrt{\mathbf{s}^t \mathbf{R}_n^{-1} \mathbf{s}}} \mathbf{R}_n^{-1} \mathbf{s} \tag{4.69}$$

Chapter 5

Control Theory

U sually, the area of control systems requires students to deal with many formulae and expressions used to compute gains, times, and parameters. Some of those formulae frequently are given without proof. In this chapter we will present a few of those proofs in the hope they will clarify the concepts and the formulae themselves.

5.1 Laplace transform

In the following sections, we will derive some properties and equations about the Laplace transform. The definition we will use is sometimes called the "causal" definition. In that definition, the Laplace transform of a function $f(t)$ is a function of s given by (frequently called complex frequency domain function)

$$F(s) = L\{f(t)\} = \int_0^\infty f(t) e^{-st} dt \qquad (5.1)$$

5.1.1   Properties

Differentiation

Using the definition, the Laplace transform for $\dot{f}(t)$ is

$$L\{\dot{f}(t)\} = \int_0^\infty \dot{f}(t)e^{-st}dt \qquad (5.2)$$

Using the technique of integration by parts (see chapter 2), we can define

$$\begin{aligned} du &= \dot{f}(t)dt \\ v &= e^{-st} \end{aligned} \qquad (5.3)$$

which, after proper differentiation and integration, changes to

$$\begin{aligned} u &= f(t) \\ dv &= -se^{-st}dt \end{aligned} \qquad (5.4)$$

Now, using the results of integration by parts, we have

$$L\{\dot{f}(t)\} = \int_0^\infty v\,du$$

$$= uv\big|_0^\infty - \int_0^\infty u\,dv \qquad (5.5)$$

$$= e^{-st}f(t)\big|_0^\infty - \int_0^\infty -se^{-st}f(t)dt$$

which leads us to

$$L\{\dot{f}(t)\} = f(0) + s\int_0^\infty e^{-st}f(t)dt \qquad (5.6)$$

We then identify the second term in the right-hand side of 5.6 as the Laplace of $f(t)$ and have ■

$$L\{\dot{f}(t)\} = f(0) + s\,F(s) \tag{5.7}$$

Integration

With a similar procedure done in the previous section, we wish to compute

$$L\left\{\int_0^t f(\tau)\mathrm{d}\tau\right\} = \int_0^\infty \int_0^t f(\tau)\mathrm{d}\tau\, e^{-st}\mathrm{d}t \tag{5.8}$$

Now, we define

$$u = \int_0^t f(\tau)\mathrm{d}\tau$$

$$dv = e^{-st}\mathrm{d}t \tag{5.9}$$

that leads to

$$du = f(t)\mathrm{d}t$$

$$v = -\frac{e^{-st}}{s} \tag{5.10}$$

Finally, we integrate by parts and get

$$L\left\{\int_0^t f(\tau)\mathrm{d}\tau\right\} = \int_0^\infty u\,dv$$

$$= uv\big|_0^\infty - \int_0^\infty v\,du \tag{5.11}$$

$$= -\frac{e^{-st}}{s}\int_0^t f(\tau)\mathrm{d}\tau\,\bigg|_0^\infty - \int_0^\infty \frac{e^{-st}}{s}f(t)\mathrm{d}t$$

with

$$\frac{e^{-st}}{s} \int_0^t f(\tau)d\tau \Bigg|_0^\infty$$

$$= \frac{e^{-\infty}}{s} \int_0^\infty f(\tau)d\tau - \frac{e^{-0}}{s} \int_0^0 f(\tau)d\tau = 0$$

(5.12)

which leads us to

$$L\left\{ \int_0^t f(\tau)d\tau \right\} = \frac{1}{s} \int_0^\infty e^{-st} f(t)dt$$

(5.13)

We then identify the second term in the right-hand side of 5.13 as the Laplace of and have ■

$$L\left\{ \int_0^t f(\tau)d\tau \right\} = \frac{F(s)}{s}$$

(5.14)

Convolution

Now, we wish to analyse the following Laplace transform

$$L\left\{ f(t) * g(t) \right\} = \int_0^\infty \int_0^\infty f(\tau)g(t-\tau)d\tau e^{-st}dt$$

(5.15)

First, we separate the integrals as

$$L\left\{ f(t) * g(t) \right\} = \int_0^\infty f(\tau) \left(\int_0^\infty g(t-\tau)e^{-st}dt \right) d\tau$$

(5.16)

Then, we perform the following change of variables

$$t' = t - \tau \qquad (5.17)$$

and get

$$L\{f(t) * g(t)\} = \int_0^\infty f(\tau) \left(\int_0^\infty g(t') e^{-s(t'+\tau)} dt' \right) d\tau$$
$$= \int_0^\infty f(\tau) \left(e^{-s\tau} \int_0^\infty g(t') e^{-st'} dt' \right) d\tau \qquad (5.18)$$

Now, we identify the inner integral as the Laplace transform for the function $g(t)$ and write

$$L\{f(t) * g(t)\} = \int_0^\infty f(\tau) G(s) e^{-s\tau} d\tau$$
$$= G(s) \int_0^\infty f(\tau) e^{-s\tau} d\tau \qquad (5.19)$$

which leads to the second Laplace transform and finally gives us ■

$$L\{f(t) * g(t)\} = G(s) F(s) \qquad (5.20)$$

Time displacement

Now, we would like to analyse the Laplace transform of a signal with a time displacement like

$$L\{f(t - t_0)\} = \int_0^\infty f(t - t_0) e^{-st} dt \qquad (5.21)$$

That turns out to be as simple as making the following change of variable

$$t' = t - t_0 \tag{5.22}$$

and we get

$$L\{f(t - t_0)\} = \int_0^\infty f(t') e^{-s(t'+t_0)} dt'$$

$$= \int_0^\infty f(t') e^{-st'} e^{-st_0} dt' \tag{5.23}$$

Now, we identify the integral as the Laplace transform for the function and write ■

$$L\{f(t - t_0)\} = e^{-st_0} F(s) \tag{5.24}$$

Frequency displacement

The signal can also be displaced in frequency, resulting in

$$F(s - s_0) = L\{f'(t)\} \tag{5.25}$$

We would like to analyse the relationship between the original signal $f(t)$ and the one that results from the frequency displace $f'(t)$. To do that we change variables in 5.25 using

$$s' = s - s_0 \tag{5.26}$$

and applying the definition of the Laplace transform as

$$F(s') = \int_0^\infty f(t) e^{-(s-s_0)t} dt$$

$$= \int_0^\infty f(t) e^{s_0 t} e^{-st} dt \tag{5.27}$$

Now, we identify the integral as the Laplace transform for the function $f(t) e^{s_0 t}$ and write ■

$$L\left\{f'(t)\right\} = F(s - s_0) \rightarrow f'(t) = e^{s_0 t} f(t) \tag{5.28}$$

Initial Value Theorem

If we multiply the Laplace transform of a signal $X(s)$ by s we get

$$sX(s) = L\{\dot{x}(t)\} \tag{5.29}$$

which is by definition equal to

$$sX(s) = \int_0^\infty \dot{x}(t)e^{-st}\mathrm{d}t \tag{5.30}$$

If we take the limit when $s \rightarrow \infty$ on both sides, the term inside the integral becomes zero almost everywhere except at $t = 0$. In this situation, the limit becomes

$$\lim_{s\to\infty} sX(s) = \int_{0^-}^{0^+} \dot{x}(t)\mathrm{d}t \tag{5.31}$$

$$= x(0^+) - x(0^-)$$

and for "causal" signals[1] we have the famous Initial Value Theorem stated as ■

$$\lim_{s\to\infty} sX(s) = x(0) \tag{5.32}$$

Final Value Theorem

In the same manner we used for the initial value, if we multiply the Laplace transform of a signal $X(s)$ by s, we get

---

[1] The integral for the Laplace transform formally has to include the 0 value, so for signals that satisfy $x(t) = 0$, $t < 0$ (t less than instead of less than or equal), we must include 0 in the integral.

$$sX(s) = \int_0^\infty \dot{x}(t)e^{-st}\mathrm{d}t \qquad (5.33)$$

If we take the limit now when $s \to 0$ on both sides, the term inside the integral becomes one everywhere. In this situation, the limit becomes

$$\lim_{s \to 0} sX(s) = \int_{0^-}^\infty \dot{x}(t)\mathrm{d}t$$
$$= x(\infty) - x(0^-) \qquad (5.34)$$

and for "causal" signals we have the famous Final Value Theorem stated as ■

$$\lim_{s \to 0} sX(s) = x(\infty) \qquad (5.35)$$

5.2   BIBO stability criteria

A very important aspect of a linear dynamical system is its stability. In this section we will investigate one of the stability criteria and derive its stability test. We will show the *bounded input bounded output* (BIBO) stability criterion. It states that, in order for a system to be considered stable it must have its output bounded whenever the input is bounded. That criterion guarantees that for all bounded input, the output also will be bounded. Formally, the criterion is stated as

$$\forall t, \quad |x(t)| < M_1 \to |y(t)| < M_2 \qquad (5.36)$$

where $x(t)$ is the input of the system, $y(t)$ its output and the constants M_1 and M_2 are finite.

At first, one would have to test all bounded inputs (computing the output) in order to assure stability in the BIBO sense. However, we are looking for a way to test that criterion directly from the model of the system (its transfer function). The goal is to derive a test that uses only the transfer function parameters and infers the stability of the system. In order to do that, we must start from a generic transfer function and (since the criterion is stated in the time domain) obtain its impulse response via inverse Laplace of the transfer function as

$$G(s) = K\frac{(s - z_1)(s - z_2)...(s - z_N)}{(s - p_1)...(s - p_M)...(s - p_1')^n...(s - p_N')^m} \to g(t)$$
(5.37)

We can notice that, in general, we have zeros and poles that might be real, complex or with multiplicity 2, 3, etc.

The criterion involves the input $x(t)$ and output $y(t)$ of the system represented by $G(s)$, so we must state that relation. Having the impulse response $g(t)$, the relationship between input and output is given by the convolution

$$y(t) = x(t) * g(t)$$

$$y(t) = \int_{-\infty}^{\infty} g(t - \tau)x(\tau)d\tau$$
(5.38)

The criterion enforces that for a bounded $x(t)$ we have a bounded $y(t)$. In another words, the integral must be finite. To ensure that, we have to find bounds for the output given bounds in the input. We can use the triangular inequality

$$A = B + C \to |A| \leqslant |B| + |C|$$
(5.39)

applied to the integral and have

$$|y(t)| \leqslant \int_{-\infty}^{\infty} |g(t - \tau)x(\tau)|\, d\tau$$
(5.40)

Having the absolute value of the product as the product of the absolute values, we have

$$|y(t)| \leqslant \int_{-\infty}^{\infty} |g(t-\tau)|\,|x(\tau)|\,d\tau \qquad (5.41)$$

We can now consider the worst case for the input that is bounded by $|x(t)| \leqslant M_1$ and write

$$|y(t)| \leqslant \int_{-\infty}^{\infty} |g(t-\tau)|\,M_1 d\tau$$

$$\qquad (5.42)$$

$$|y(t)| \leqslant M_1 \int_{-\infty}^{\infty} |g(t-\tau)|\,d\tau$$

since all values of $|x(t)|$ that are smaller than M_1 will only decrease the final value of the integral. In summary, considering a bounded input, in order to get a bounded output, we must ensure that the integral

$$\int_{-\infty}^{\infty} |g(\tau)|\,d\tau \leqslant M \qquad (5.43)$$

converges.

Although we now have a stability test that depends only on the transfer function parameters (impulse response $g(t)$), this test is still not easy to perform. The integral in 5.43 is frequently very hard to compute, so an alternative method is needed. We will use two conditions that are necessary for equation 5.43 to converge. They are

$$|g(t)| < \infty$$
$$\lim_{t \to \infty} |g(t)| = 0 \qquad (5.44)$$

The first one ensures that no value of the impulse response is unbounded (ensuring that the integral won't be either). The second ensures that there is no residue that might accumulate in the integral and unbound the integral.

To test those conditions, we must write explicitly an expression for the impulse response $g(t)$. To do that, we expand the transfer function in partial fractions like

$$G(s) = \frac{A_1}{s - p_1} + \ldots + \frac{A_M}{s - p_M} + \frac{B_1}{s - p_1'} + \ldots +$$
$$+ \frac{B_n}{(s - p_1')^n} + \ldots + \frac{C_1}{s - p_2'} + \ldots + \frac{C_m}{(s - p_2')^m} + \ldots \tag{5.45}$$

which leads to the following impulse response

$$g(t) = k_1 e^{p_1 t} + \ldots + k_M e^{p_M t} +$$
$$+ k_1' e^{p_1' t} + t k_2' e^{p_1' t} + \ldots + t^{n-1} k_n' e^{p_1' t} + \tag{5.46}$$
$$+ k_1'' e^{p_2' t} + t k_2'' e^{p_2' t} + \ldots + t^{m-1} k_m'' e^{p_2' t} + \ldots$$

In order to have conditions in 5.44 met, each term of 5.46 must obey 5.44. Hence, we must analyse all the possible cases for the terms in a linear time invariant system. We have basically four situations:

- $k e^{p t}$: In this case we have a real pole p. For this case, in order for the term to agree with 5.44 we have to have $p < 0$ (so the exponential will decrease)

- $k e^{p t} + k e^{\bar{p} t}$: In this case we have a pair of complex conjugate poles p and $\bar{p}$. For this case, to agree with 5.44 we have to have $\text{Re}\{p\} < 0$ (so the exponential will oscillate but decrease)

- $k t^n e^{p t}$: Here the real poles have multiplicity $n - 1$. In this case, even though the power in t grows, this growth is overcome by the exponential if we have $p < 0$. Hence, we keep the same condition as the first case.

- $k t^n (e^{p t} + k e^{\bar{p} t})$: In the same situation as the previous case, the power in t (multiplicity of complex poles) demands that we have $\text{Re}\{p\} < 0$ as the condition for convergence.

Finally, we use the fact that, for a real number p, $\text{Re}\{p\} = p$ and synthesise the whole test as ■

$$\text{Re}\{p_i\} < 0 \tag{5.47}$$

where p_i represents all poles in the transfer function. That leads to the famous stability region for linear time invariant systems as the left half-plane in the complex plane for the poles (figura 5.1).

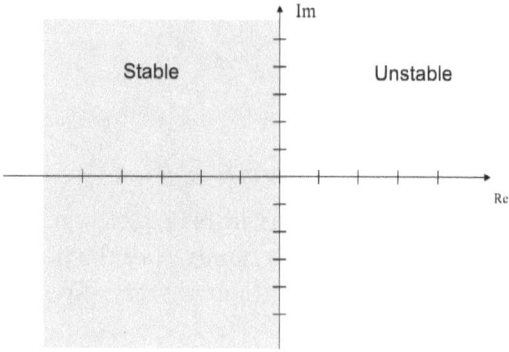

Figure 5.1: Stability region for the poles of dynamical system.

5.3 Z-transform

The Z-transform is the standard transform used to analyse discrete systems in control theory (as the Laplace transform is to analyse continuous time systems). The idea behind those transforms is that they would "invert" the operation present in the time domain model of a system. In the case of continuous systems, the Laplace transform (by being an integral transform) would invert the derivatives present in a differential equation and, in fact, turn them into algebraic equations. The idea behind the Z-transform is the same, but instead of inverting a derivative, it would invert a "difference" in a difference equation (that is the standard model for discrete systems). So, by intuition, one would expect that the Z-transform would be a "summation" transform (and in fact it is). In order to derive a form that has the same properties as the Laplace transform for continuous systems, we will proceed with the steps given in the discretisation of a continuous time signal.

The first thing to do is to represent the discrete version of a signal $x(t)$ by a signal that exists only at discrete times $k\,T$ (with $k = 0; 1; 2; \ldots$ and T being the sampling time). This representation can be written as follows

$$\hat{x}(t) = \sum_{k=0}^{\infty} x(t)\delta(t - kT) \qquad (5.48)$$

where $\delta(.)$ is the Dirac delta function. The signal $\hat{x}(t)$ is still a continuous time signal, but it exists only at the instants $k\,T$. Because of that we can change the argument of the signal $x(t)$ by $x(k\,T)$ without losing information. Moreover, we can use Oppenheim's[2] notation and write

$$\hat{x}(t) = \sum_{k=0}^{\infty} x[k]\delta(t - kT) \qquad (5.49)$$

If we wish to represent it in the complex frequency domain, we must use the Laplace transform as

$$\mathcal{L}\{\hat{x}(t)\} = \mathcal{L}\left\{\sum_{k=0}^{\infty} x[k]\delta(t - kT)\right\} \qquad (5.50)$$

As the Laplace transform is a linear operator, we can write

$$\hat{X}(s) = \sum_{k=0}^{\infty} x[k]\mathcal{L}\{\delta(t - kT)\} \qquad (5.51)$$

Now, taking the Laplace of the translated Dirac delta function, we obtain

---

[2]This notation was proposed by Oppenheim and is used to distinguish a continuous signal from a discrete one by the use of square brackets instead of parentheses.

$$\hat{X}(s) = \sum_{k=0}^{\infty} x[k]e^{-kTs} \tag{5.52}$$

Since T is considered constant during all time in the analysis of discrete systems, we can make a change of variable that rewrites the complex frequency as $z = e^{sT}$ and we get

$$X(z) = \sum_{k=0}^{\infty} x[k]z^{-k} \tag{5.53}$$

Which is the Z-transform of the sequence $X[k]$ ■.

5.3.1   Properties

Summation

Consider the signal $y[n]$ as

$$y[n] = \sum_{i=0}^{n} x[i] \tag{5.54}$$

We wish to express the Z-transform $Y(z)$ as a function of the the Z-transform of the signal $x[n]$ ($X(z)$). To do that we could write $y[n-1]$

$$y[n-1] = \sum_{i=0}^{n-1} x[i] \tag{5.55}$$

and subtract from equation 5.54 leading to

$$y[n] - y[n-1] = \sum_{i=0}^{n} x[i] - \sum_{i=0}^{n-1} x[i] \tag{5.56}$$

$$= x[n]$$

Now, taking the Z-transform, we have

$$Y(z) - z^{-1}Y(z) = X(z) \tag{5.57}$$

and

hence ■

$$Y(z) = \frac{X(z)}{1 - z^{-1}} \tag{5.58}$$

Differentiation in Z

Consider the signal in the frequency domain $X(z)$. By definition

$$X(z) = \sum_{n=0}^{\infty} x[n]z^{-n} \tag{5.59}$$

If we differentiate with respect to z, we have

$$\frac{\mathrm{d}X(z)}{\mathrm{d}z} = \sum_{n=0}^{\infty} -nx[n]z^{-n-1} \tag{5.60}$$

which we can rewrite as

$$\frac{\mathrm{d}X(z)}{\mathrm{d}z} = z^{-1} \sum_{n=0}^{\infty} -nx[n]z^{-n} \tag{5.61}$$

That leads to

$$\frac{\mathrm{d}X(z)}{\mathrm{d}z} = Z\left\{-nx[n-1]\right\} \tag{5.62}$$

Initial Value Theorem

If we have the Z-transform for a variable $x[k]$ (say $X(z)$), we can obtain the value of $x[0]$ directly from the Z-Transform by applying the limit

$$x[0] = \lim_{z \to \infty} X(z) \tag{5.63}$$

To show that, we simple plug in the definition of the Z-transform and obtain

$$\lim_{z \to \infty} (z - 1) \sum_{k=0}^{\infty} x[k]z^{-k} =$$

$$\lim_{z \to \infty} x[0] + x[1]z^{-1} + x[2]z^{-2} + x[3]z^{-3} + \ldots \tag{5.64}$$

If we compute the limit, all the terms will be zero except the first one

$$\lim_{z \to \infty} X(z) = x[0] + x[1] \cdot 0 + x[2] \cdot 0 + x[3] \cdot 0 + \cdots = x[0] \tag{5.65}$$

as we stated before■.

Final Value Theorem

The Initial Value Theorem makes it possible to obtain the "initial" value of the sequence in the time domain ($x[0]$), using the "final" value in the frequency domain ($X(z)$). Interestingly, it is possible to do the opposite and obtain the "final" value for $x[k]$ by using the "initial" value for $X(z)$. That is called the Final Value Theorem. Formally, we have

$$\lim_{k \to \infty} x[k] = \lim_{z \to 1} (z - 1)X(z) \tag{5.66}$$

Plugging in the definition for the Z-transform, we have

$$\lim_{z \to 1} (z - 1)X(z) = \lim_{z \to 1} (z - 1) \sum_{k=0}^{\infty} x[k]z^{-k} \tag{5.67}$$

Expanding the right-hand side we get

$$\lim_{z \to 1} \left(\sum_{k=0}^{\infty} x[k]z^{-k+1} - \sum_{k=0}^{\infty} x[k]z^{-k} \right)$$

$$= \lim_{z \to 1} \left(x[0]z + \sum_{k=1}^{\infty} x[k]z^{-k} - \sum_{k=0}^{\infty} x[k]z^{-k} \right) \tag{5.68}$$

Now, performing the change of variable given by $k = k' + 1$, we obtain

$$\lim_{z \to 1} \left(x[0]z + \sum_{k'=0}^{\infty} x[k'+1]z^{-k'} - \sum_{k=0}^{\infty} x[k]z^{-k} \right) \tag{5.69}$$

We can now write the summation as a limit as

$$\lim_{z \to 1} \left(\lim_{M \to \infty} \left(x[0]z + \sum_{k'=0}^{M} x[k'+1]z^{-k'} - \sum_{k=0}^{M} x[k]z^{-k} \right) \right) \tag{5.70}$$

Expanding the summations, we get

$$\lim_{z \to 1} \left(\lim_{M \to \infty} \left(\begin{array}{l} x[0]z + x[1] + x[2]z^{-1} + \ldots + x[M+1]z^{-M} - \\ -x[0] - x[1]z^{-1} - x[2]z^{-2} - \ldots - x[M]z^{-M} \end{array} \right) \right) \tag{5.71}$$

which, grouping the terms with same $x[k]$, leads us to

$$\lim_{z \to 1} \left(\lim_{M \to \infty} \left(\begin{array}{l} x[0](z-1) + x[1](1-z^{-1}) + x[2](z^{-1}-z^{-2}) + \ldots \\ +x[M](z^{-M+1} - z^{-M}) + x[M+1]z^{-M} \end{array} \right) \right) \tag{5.72}$$

Now, as $z \to 1$, all the terms goes to zero except the last one, leading us to

$$\lim_{M \to \infty} \left(x[M+1]1^{-M} \right) = \lim_{M \to \infty} x[M] \tag{5.73}$$

which is equal to the left-hand side of 5.66 ∎.

5.3.2 Convolution

The discrete convolution of two signals in the time domain is defined as

$$x_1[k] * x_2[k] = \sum_{i=0}^{\infty} x_1[i]x_2[k - i] \qquad (5.74)$$

If we apply the Z-transform to that convolution as

$$Z\left\{x_1[k] * x_2[k]\right\} = Z\left\{\sum_{i=0}^{\infty} x_1[i]x_2[k - i]\right\} \qquad (5.75)$$

we have the following double summation

$$Z\left\{\sum_{i=0}^{\infty} x_1[i]x_2[k - i]\right\} = \sum_{k=0}^{\infty}\sum_{i=0}^{\infty} x_1[i]x_2[k - i]z^{-k} \qquad (5.76)$$

We can exchange the summation and since the first term $x_1[i]$ does not depend on k, we can move out from the first summation and have

$$Z\left\{x_1[k] * x_2[k]\right\} = \sum_{i=0}^{\infty} x_1[i]\sum_{k=0}^{\infty} x_2[k - i]z^{-k} \qquad (5.77)$$

Now, we identify the second summation as the Z-transform of the signal $x_2[k]$ translated by i and (using the property of time translation) we have

$$Z\left\{x_1[k] * x_2[k]\right\} = \sum_{i=0}^{\infty} x_1[i]X_2(z)z^{-i}$$
$$X_2(z)\sum_{i=0}^{\infty} x_1[i]z^{-i} \qquad (5.78)$$

Now, we identify the next summation as the Z-transform of the signal $x_1[k]$ and finally have ∎.

$$Z\left\{x_1[k] * x_2[k]\right\} = X_1(z)X_2(z) \qquad (5.79)$$

5.3.3   Some Z-transform examples

Discrete impulse

The discrete impulse is defined as

$$p[n] = \begin{cases} 1, & n = 0 \\ 0, & n \neq 0 \end{cases} \tag{5.80}$$

so, as trivial as it is ■,

$$Z\{p[n]\} = \sum_{n=0}^{\infty} p[n] z^{-n}$$
$$1 \cdot z^0 + 0 \cdot z^{-1} + 0 \cdot z^{-2} + \dots \tag{5.81}$$
$$= 1$$

Exponential

Let

$$y[n] = a^n 1[n] \tag{5.82}$$

where $1[n]$ is the step function and $|a| < 1$. The Z-transform can be calculated using the definition by

$$Y(z) = \sum_{n=0}^{\infty} a^n z^{-n} \tag{5.83}$$

Expanding the summation, we have

$$Y(z) = 1 + az^{-1} + \left(az^{-1}\right)^2 + \left(az^{-1}\right)^3 + \dots \tag{5.84}$$

which is a geometric progression with common ratio equal to $a\,z^{-1}$. So the summation will lead to (see section 7.4.2)

$$Y(z) = \frac{1}{1 - az^{-1}} \tag{5.85}$$

$na^n 1[n]$

Applying the Z-transform definition to the signal, we have

$$\sum_{n=0}^{\infty} na^n z^{-n} \tag{5.86}$$

Using the result from section 7.4.4, we can rewrite the summation as

$$\sum_{n=0}^{\infty} n\left(az^{-1}\right)^n \tag{5.87}$$

That leads to

$$\frac{az^{-1}}{\left(1 - az^{-1}\right)^2} \tag{5.88}$$

$\sin\left(\varpi_0 n + \theta\right)$

If we write the expression with its Euler equivalent (see section 7.1.1), we have

$$\sin\left(\varpi_0 n + \theta\right) = \frac{e^{j\varpi_0 n + j\theta} - e^{-j\varpi_0 n - j\theta}}{2j} \tag{5.89}$$

We can express the two terms as

$$\sin\left(\varpi_0 n + \theta\right) = \frac{e^{j\theta}}{2j}\left(e^{j\varpi_0}\right)^n - \frac{e^{-j\theta}}{2j}\left(e^{-j\varpi_0}\right)^n \tag{5.90}$$

Now, applying the Z-transform, we have

$$Z\left\{\sin\left(\varpi_0 n + \theta\right)\right\} =$$

$$\sum_{n=0}^{\infty} \frac{e^{j\theta}}{2j}\left(e^{j\varpi_0}\right)^n z^{-n} - \sum_{n=0}^{\infty} \frac{e^{-j\theta}}{2j}\left(e^{-j\varpi_0}\right)^n z^{-n} \tag{5.91}$$

which can be rewritten as

$$Z\left\{\sin\left(\varpi_0 n + \theta\right)\right\} =$$

$$\frac{e^{j\theta}}{2j} \sum_{n=0}^{\infty} \left(e^{j\varpi_0} z^{-1}\right)^n - \frac{e^{-j\theta}}{2j} \sum_{n=0}^{\infty} \left(e^{-j\varpi_0} z^{-1}\right)^n \qquad (5.92)$$

Now, identifying the two geometric progressions, we have

$$Z\left\{\sin\left(\varpi_0 n + \theta\right)\right\} =$$

$$\frac{e^{j\theta}}{2j} \frac{1}{1 - e^{j\varpi_0} z^{-1}} - \frac{e^{-j\theta}}{2j} \frac{1}{1 - e^{-j\varpi_0} z^{-1}} \qquad (5.93)$$

which can be rewritten as

$$Z\left\{\sin\left(\varpi_0 n + \theta\right)\right\} =$$

$$\frac{2j e^{j\theta}(1 - e^{-j\varpi_0} z^{-1}) - 2j e^{-j\theta}(1 - e^{j\varpi_0} z^{-1})}{-4(1 - e^{j\varpi_0} z^{-1})(1 - e^{-j\varpi_0} z^{-1})}$$

$$= \frac{\frac{1}{2j}\left(e^{j\theta} - e^{-j\theta}\right) + \frac{1}{2j}\left(e^{j(\varpi_0 - \theta)} - e^{-j(\varpi_0 - \theta)}\right) z^{-1}}{\left(1 + 2\frac{1}{2}\left(e^{j\varpi_0} - e^{-j\varpi_0}\right) z^{-1} + z^{-2}\right)} \qquad (5.94)$$

That finally leads to

$$Z\left\{\sin\left(\varpi_0 n + \theta\right)\right\} = \frac{\sin(\theta) + \sin\left(\varpi_0 - \theta\right) z^{-1}}{1 + 2\cos(\varpi_0) z^{-1} + z^{-2}} \qquad (5.95)$$

5.4 Stability criteria

Here we are going to derive the condition for bounded input bounded output (BIBO) stability for discrete systems. We start with the formal definition for BIBO stability (as done in the continuous case) as

$$\forall k, \ |u[k]| \leqslant M_1 \Rightarrow |y[k]| \leqslant M_2 \qquad (5.96)$$

Since the definition is formally stated in the time domain, we need to write the output also in the time domain. Hence, we write

$$y[k] = h[k] * u[k] \tag{5.97}$$

that translates to

$$y[k] = \sum_{i=0}^{\infty} h[i]u[k-i] \tag{5.98}$$

We follow the same reasoning as in the continuous case (see section 5.2) and use the triangular inequality (together with the fact that $|AB| = |A||B|$) to write

$$|y[k]| \leqslant \sum_{i=0}^{\infty} |h[i]||u[k-i]| \tag{5.99}$$

Now, we also analyse the worst case for the input and have

$$|u[k-i]| \leqslant M_1 \rightarrow \max |u[k-i]| = M_1 \tag{5.100}$$

So, we can write

$$\begin{aligned} |y[k]| &\leqslant \sum_{i=0}^{\infty} |h[i]| M_1 \\ &\leqslant M_1 \sum_{i=0}^{\infty} |h[i]| \end{aligned} \tag{5.101}$$

Hence, for a finite output ($|y[k]| \leqslant M_2$), we have

$$\sum_{i=0}^{\infty} |h[i]| < \infty \tag{5.102}$$

As in the continuous case, this definition, although generic enough to test stability independent of the input, is very hard to compute in general. To overcome this problem we write $h[k]$ as the inverse Z-transform of (expanding in partial fractions)

$$H(z) = \frac{N(z)}{D(z)} = \frac{A_1}{1 - z_1 z^{-1}} + \frac{A_2}{1 - z_2 z^{-1}} + \ldots + \frac{A_n}{1 - z_n z^{-1}} \tag{5.103}$$

leading to [3]

$$h[k] = A_1 z_1^k + A_2 z_2^k + \dots + A_n z_n^k \qquad (5.104)$$

We then need the following summation to converge

$$\sum_{k=0}^{\infty} h[k] = A_1 \sum_{k=0}^{\infty} z_1^k + A_2 \sum_{k=0}^{\infty} z_2^k + \dots + A_n \sum_{k=0}^{\infty} z_n^k \qquad (5.105)$$

which implies ■

$$\sum_{k=0}^{\infty} z_i^k < \infty \rightarrow |z_i| < 1 \qquad (5.106)$$

That result indicates the famous region of convergence for LTI discrete systems as the unity circle (values with modulus less than one) as in the figure 5.2.

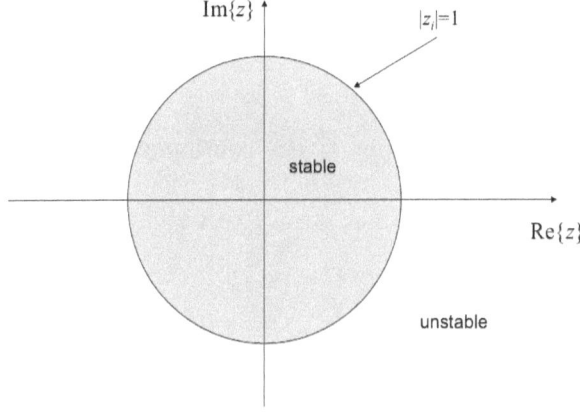

Figure 5.2: Stability region for the poles of dynamical system (discrete case).

---

[3]It is important to notice that z_i are roots that might be complex or repeated. In such cases, the same reasoning used in section 5.2 holds and all the cases for complex and repeated roots can be stated as the case with a single, non repeating, root.

5.5   Discretisation approximations

In this section we will derive the approximations used to discretise continuous systems by changing from the s Laplace variable to the z Z-transform variable. For each case we will derive the region of stability as well. As we know, some of the approximations do not map completely the left half-plane to the unity circle.

5.5.1   ZOH

The first discretisation method we will derive will actually not be an approximation but rather an exact method. This method is often called "step invariant" and the goal is to make a discretisation that preserves the value of a step response of the system in the sampled times. The idea is that, since the ZOH[4] of a step signal is identical to its continuous source, we will pick its response to match the discrete values to the continuous values at the sampling points in time. Formally, we have

$$y_c(h\,k) = y_d[k] \qquad (5.107)$$

where y_d is the output of the continuous system, h is the sampling time and y_d is the output of the discretised system. To use the step as input, it is easy to write the relationship in the frequency domain as

$$L^{-1}\{Y_c(s)\}\big|_{t=hk} = Z^{-1}\{Y(z)\} \qquad (5.108)$$

Computing the step response for each case, we have

$$
\begin{aligned}
L^{-1}\{Y_c(s)\}\big|_{t=hk} &= L^{-1}\left\{\frac{1}{s}C_c(s)\right\}\bigg|_{t=hk} \\
Z^{-1}\{Y(z)\} &= Z^{-1}\left\{\frac{z}{z-1}C_d(z)\right\}
\end{aligned}
\qquad (5.109)
$$

---

[4]ZOH stands for zero order holder and is a very common abbreviation used in discrete systems.

where $C_s(s)$ and $C_d(z)$ are the transfer function of the system (continuous and discrete respectively). Now, solving for $C_d(z)$, we have

$$C_d(z) = \frac{z-1}{z} Z \left\{ L^{-1} \left\{ \frac{1}{s} C_c(s) \right\} \Big|_{t=hk} \right\} \tag{5.110}$$

which computes the discrete transfer function of a system given the continuous one and the sampling time ■

Stability analysis

If we write equation 5.110 using partial fractions for the continuous transfer function, we have

$$C_d(z) = \frac{z-1}{z} Z \left\{ L^{-1} \left\{ \frac{A_0}{s} + \frac{A_1}{s+p_1} + \ldots + \frac{A_n}{s+p_n} \right\} \Big|_{t=hk} \right\} \tag{5.111}$$

which develops to

$$C_d(z) = \frac{z-1}{z} Z \left\{ \left(A_0 + A_1 e^{-p_1 t} + \ldots + A_n e^{-p_n t} \right) u(t) \big|_{t=hk} \right\}$$

$$= \frac{z-1}{z} Z \left\{ \left(A_0 + A_1 e^{-p_1 hk} + \ldots + A_n e^{-p_n hk} \right) u(hk) \right\} \tag{5.112}$$

and finally

$$C_d(z) = \frac{z-1}{z} \left(\frac{A_0 z}{z-1} + \frac{A_1 z}{z - e^{-p_1 h}} + \ldots + \frac{A_n z}{z - e^{-p_n h}} \right)$$

$$C_d(z) = A_0 + \frac{(z-1)A_1}{(z - e^{-p_1 h})} + \ldots + \frac{(z-1)A_n}{(z - e^{-p_n h})} \tag{5.113}$$

The impulse response for the discrete transfer function is finally

$$C_d[k] = k_1 p[k] + k_2 1[k] + k_3 e^{-p_1 h(k-1)} + \ldots + k_{n+2} e^{-p_n h(k-1)} \tag{5.114}$$

For that impulse response to converge (and therefore lead to a stable transfer function), we must have

$$\text{Re}\{p_i\} < 0 \Rightarrow Ke^{-p_i h(k-1)} \to 0 \qquad (5.115)$$

Here, the values of the p_is are the poles of the original continuous system. The conclusion is that for the discrete version of the transfer function to be stable we must have $\text{Re}\{p_i\} < 0$, which is the same condition for the original system. Hence, stable continuous systems will map to stable discrete ones and unstable continuous systems will also map to unstable continuous systems, thus making a correct mapping.

5.5.2   Forward difference

One of the methods to approximate the discrete transfer function given a continuous one is by mapping the derivative in both domains (continuous and discrete). That is done by mapping the differentiation operation in the Laplace transform as well as in the Z-transform as

$$\dot{y}(t) \leftrightarrow sY(s) \qquad (5.116)$$

Approximating the derivative, we have

$$\frac{y(t+h) - y(t)}{h} \approx \dot{y}(t) \qquad (5.117)$$

Now, for the discrete case, we can write

$$\frac{y[k+1] - y[k]}{h} \leftrightarrow sY(s) \qquad (5.118)$$

Which, after Z-transform, leads to

$$Y(z)\frac{z-1}{h} \leftrightarrow sY(s) \qquad (5.119)$$

Equation 5.119 tells us that in order to approximate a continuous transfer function by a discrete one we must simply change the variable s by $\frac{z-1}{h}$, representing the approximation of the differentiation operation. Thus we have ■

$$C_d(z) = C_c(s)|_{s=\frac{z-1}{h}} \tag{5.120}$$

Stability analysis

In order to compare the stability of both transfer functions ($C_c(s)$ and $C_d(z)$), we start with the mapping

$$s = \frac{z-1}{h} \tag{5.121}$$

and write

$$z = s\,h + 1 \tag{5.122}$$

Now, we consider $s = \sigma + j\omega$ and take the modulus of the variable z so we can test for unity

$$|z| = \sqrt{(1 + h\sigma)^2 + (h\varpi)^2} \tag{5.123}$$

For stability in the Z domain, we have

$$|z| < 1 \Rightarrow \sqrt{(1 + h\sigma)^2 + (h\varpi)^2} < 1 \tag{5.124}$$

which leads to

$$(1 + h\sigma)^2 + (h\varpi)^2 < 1 \tag{5.125}$$

If we have a very small value for the sampling time h, we need σ to be negative in order for 5.125 to hold. That is in accordance with stability in the continuous case (since $\sigma < 0$ is the stability condition). However, if h is not small enough, any value of ω can make the left-hand side of 5.125 greater than one, leading to an unstable discrete system (even though the continuous one is stable).

5.5.3   Backward difference

Another method to approximate the derivative in both domains (continuous and discrete) is done by mapping the differentiation operation in the Laplace transform with an approximation that takes the past value for the continuous function as

$$\frac{y(t) - y(t - h)}{h} \approx \dot{y}(t) \tag{5.126}$$

which for the discrete case we can write

$$\frac{y[k] - y[k - 1]}{h} \leftrightarrow sY(s) \tag{5.127}$$

And, after Z-transform, leads to

$$Y(z)\frac{1 - z^{-1}}{h} \leftrightarrow sY(s) \tag{5.128}$$

Now, the mapping is done by changing the variable s by $\frac{1-z^{-1}}{h}$, representing the approximation of the differentiation operation. Thus we have ▮

$$C_d(z) = C_c(s)|_{s=\frac{z-1}{zh}} \tag{5.129}$$

Stability analysis

Now, in order to compare the stability of both transfer functions ($C_c(s)$ and $C_d(z)$), we start with the mapping as we did in the forward difference

$$s = \frac{z - 1}{zh} \tag{5.130}$$

and write

$$z = \frac{1}{1 - sh} \tag{5.131}$$

Now, we consider the same representation for the Laplace variable $s = \sigma + jw$ and take the modulus of the variable z so we can test for unity

$$|z| = \frac{1}{|1 - h\sigma - jhw|} = \frac{1}{\sqrt{(1 - h\sigma)^2 + (hw)^2}} \qquad (5.132)$$

For stability in the Z domain, we have

$$|z| < 1 \rightarrow \frac{1}{\sqrt{(1 - h\sigma)^2 + (hw)^2}} < 1 \qquad (5.133)$$

which leads to

$$(1 - h\sigma)^2 + (hw)^2 > 1 \qquad (5.134)$$

Now, a situation that is the opposite of the forward difference occurs. We can have an unstable system in the continuous case (with a $\sigma > 0$) and compensate for the difference with some value of w that makes the left-hand side of 5.134 larger than one.

5.5.4 Bilinear approximation

Another way to find an approximate relation between the continuous transfer function and the discrete counterpart is to approximate the integral rather than the derivative. We start with the relation between time and frequency as we did in the previous sections

$$\int y(t)dt \leftrightarrow \frac{1}{s}Y(s) \qquad (5.135)$$

Now, we use the first Simpson Rule to approximate the integral and write

$$\frac{h}{2}\sum_{i=0}^{k-1}(y[i]+y[i+1]) \approx \int y(t)\mathrm{d}t \qquad (5.136)$$

Now, we use the Z-transform for the approximation and get

$$\frac{h}{2}\frac{z+1}{z-1}Y(z) \leftrightarrow \frac{1}{s}Y(s) \qquad (5.137)$$

leading to the mapping

$$C_d(z) = C_c(s)|_{s=\frac{2}{h}\frac{z-1}{z+1}} \qquad (5.138)$$

Stability analysis

To analyse stability we proceed as we did in the previous cases and start with the mapping

$$s = \frac{2}{h}\frac{1-z^{-1}}{1+z^{-1}} \qquad (5.139)$$

We then isolate z as

$$z = \frac{1+\frac{h}{2}s}{1-\frac{h}{2}s} \qquad (5.140)$$

and write its modulus as

$$|z| = \sqrt{\frac{\left(1+\frac{h}{2}\sigma\right)^2 + \left(\frac{h}{2}\varpi\right)^2}{\left(1-\frac{h}{2}\sigma\right)^2 + \left(\frac{h}{2}\varpi\right)^2}} \qquad (5.141)$$

We can see that stability in the discrete case ($|z| < 1$) depends only on σ since the term in w appears the same in the numerator and denominator of 5.141. Moreover, we notice that regardless of the value of h, negative values for *sigma* will make the denominator greater than the numerator and vice versa. Thus, we have the following mapping

$$|z| < 1 \Rightarrow \sigma < 0$$
$$|z| = 1 \Rightarrow \sigma = 0 \qquad (5.142)$$
$$|z| > 1 \Rightarrow \sigma > 0$$

which is a perfect mapping in terms of stability between continuous and discrete systems.

5.6 Digital PID

In this section we will derive an expression for the implementation of the discrete PID controller. The expression will be in the form of a difference equation suitable for implementation in most programming languages or devices. The idea is to write the control signal $u[k]$ as a function of past values of the error $e[k - i]$ and/or past values of the control signal itself $u[k - i]$. To do that we will start with the continuous form of the PID controller. We write the error signal as a summation of parts that are proportional to the error itself, proportional to the integral of the error and to the derivative of the error.

$$u(t) = K_p e(t) + K_i \int e(t) \mathrm{d}t + K_d \dot{e}(t) \qquad (5.143)$$

The proportionality constants K_p, K_i and K_d are the control parameters. Writing the approximation for each part, we have

$$u[k] = K_p e[k] + K_i h \sum_{i}^{k-1} e[i] + K_d \frac{e[k] - e[k-1]}{h} \qquad (5.144)$$

This expression is clearly not suitable for implementation because of the summation (approximation for the integral). It would require storing the past values for the error which is impossible. One alternative is to accumulate the integral part and use it in the expression. However, we can have a more elegant solution by writing the control signal $u[k - 1]$ as

$$u[k-1] = K_p e[k-1] + K_i h \sum_{i}^{k-2} e[i] + K_d \frac{e[k-1] - e[k-2]}{h}$$

$$(5.145)$$

and subtracting it from both sides of 5.144 as

$$u[k] - u[k-1] =$$

$$K_p e[k] + K_i h \sum_{i}^{k-1} e[i] + K_d \frac{e[k] - e[k-1]}{h}$$

$$-K_p e[k-1] - K_i h \sum_{i}^{k-2} e[i] - K_d \frac{e[k-1] - e[k-2]}{h}$$

$$(5.146)$$

which leads to a single expression for the control signal ∎.

$$u[k] = u[k-1] +$$
$$K_p \left(e[k] - e[k-1] \right) +$$
$$K_i h e[k-1] +$$
$$K_d \frac{e[k] - 2e[k-1] + e[k-2]}{h}$$

$$(5.147)$$

5.7   Observability and controllability

It is common to find books presenting the matrices that inform about the observability and controllability for a given state space representation of a linear system. To test for controllability or observability one must compute the rank of the respective matrix and test whether this rank is equal to the number of states n. In the following sections we will derive this procedure and justify why one must do that kind of test. We will divide the derivation for the cases of continuous and discrete systems.

5.7.1 Observability

To analyse the observability of a given representation, we must leave the system free of inputs or at least leave it constant (otherwise one could change the inputs to make the system behave like another one) and observe its dynamics. A representation is said observable if, measuring only its outputs, one is able to estimate the values of the states in the system currently being represented. "Measure the output" means that one can access the output values and their derivatives (as many as necessary).

Continuous systems

Let the state space representation of a linear system be given by

$$\dot{\mathbf{x}}(t) = \mathbf{A}\dot{\mathbf{x}}(t)$$

$$\begin{bmatrix} \dot{x}_1(t) \\ \vdots \\ \dot{x}_n(t) \end{bmatrix} = \mathbf{A} \begin{bmatrix} x_1(t) \\ \vdots \\ x_n(t) \end{bmatrix} \qquad (5.148)$$

Since the definition talks about derivatives of the output, we can differentiate both sides and have

$$\begin{bmatrix} \ddot{x}_1(t) \\ \vdots \\ \ddot{x}_n(t) \end{bmatrix} = \mathbf{A} \begin{bmatrix} \dot{x}_1(t) \\ \vdots \\ \dot{x}_n(t) \end{bmatrix} \qquad (5.149)$$

Substituting the derivatives of 5.148 into the right side of 5.149 we get

$$\begin{bmatrix} \ddot{x}_1(t) \\ \vdots \\ \ddot{x}_n(t) \end{bmatrix} = \mathbf{A}^2 \begin{bmatrix} x_1(t) \\ \vdots \\ x_n(t) \end{bmatrix} \qquad (5.150)$$

If we differentiate again and again, we will find

$$
\begin{bmatrix} x_1^{(n-1)}(t) \\ \vdots \\ x_n^{(n-1)}(t) \end{bmatrix} = \mathbf{A}^{n-1} \begin{bmatrix} x_1(t) \\ \vdots \\ x_n(t) \end{bmatrix} \qquad (5.151)
$$

Now, consider the output of the system

$$
\mathbf{y}(t) = \mathbf{C} \begin{bmatrix} x_1(t) \\ \vdots \\ x_n(t) \end{bmatrix} \qquad (5.152)
$$

and differentiate both sides. We then have

$$
\dot{\mathbf{y}}(t) = \mathbf{C} \begin{bmatrix} \dot{x}_1(t) \\ \vdots \\ \dot{x}_n(t) \end{bmatrix} = \mathbf{CA} \begin{bmatrix} x_1(t) \\ \vdots \\ x_n(t) \end{bmatrix} \qquad (5.153)
$$

that differentiating $n - 1$ times gives us

$$
\mathbf{y}^{(n-1)}(t) = \mathbf{CA}^{n-1} \begin{bmatrix} x_1(t) \\ \vdots \\ x_n(t) \end{bmatrix} \qquad (5.154)
$$

Following the concept of observability, if for a system of order n, we measure the output and its derivatives in a given time t, we must be able to compute the states $\mathbf{x}(t)$. Therefore, for it to be controllable, we must be able to solve the following system of equations

$$\mathbf{y}(t) = \mathbf{C} \begin{bmatrix} x_1(t) \\ \vdots \\ x_n(t) \end{bmatrix}$$

$$\dot{\mathbf{y}}(t) = \mathbf{CA} \begin{bmatrix} x_1(t) \\ \vdots \\ x_n(t) \end{bmatrix} \qquad (5.155)$$

$$\vdots$$

$$\mathbf{y}^{(n-1)}(t) = \mathbf{CA}^{n-1} \begin{bmatrix} x_1(t) \\ \vdots \\ x_n(t) \end{bmatrix}$$

Writing 5.155 in matrix form, we have

$$\begin{bmatrix} \mathbf{C} \\ \mathbf{CA} \\ \vdots \\ \mathbf{CA}^{n-1} \end{bmatrix} \begin{bmatrix} x_1(t) \\ x_2(t) \\ \vdots \\ x_n(t) \end{bmatrix} = \begin{bmatrix} \mathbf{y}(t) \\ \dot{\mathbf{y}}(t) \\ \vdots \\ \mathbf{y}^{(n-1)}(t) \end{bmatrix} \qquad (5.156)$$

where we define the observability matrix as

$$\mathbf{O} = \begin{bmatrix} \mathbf{C} \\ \mathbf{CA} \\ \vdots \\ \mathbf{CA}^{n-1} \end{bmatrix} \qquad (5.157)$$

Therefore, for that system to have a solution, the matrix $\mathbf{O}$ must be non-singular or (if it is nor square) have rank equal to n ■.

So, the rank test is nothing but a test for solvability of a linear system. That is why we use the determinant when the observability matrix is square.

Discrete time systems

As in the case of continuous systems, for a representation of a discrete system to be observable, we must be able to compute its states given its outputs. Now, instead of measuring the output and its derivatives we measure the output at different instants $k, k+1, \ldots, k+n-1$ where n is the order of the system (number of states).

Consider the discrete representation given by (we again ignore the input because we want no disturbances in the dynamic of the system)

$$
\begin{bmatrix} x_1[k+1] \\ \vdots \\ x_n[k+1] \end{bmatrix} = \Phi \begin{bmatrix} x_1[k] \\ \vdots \\ x_n[k] \end{bmatrix} \tag{5.158}
$$

if we observe one step ahead, we get

$$
\begin{bmatrix} x_1[k+2] \\ \vdots \\ x_n[k+2] \end{bmatrix} = \Phi \begin{bmatrix} x_1[k+1] \\ \vdots \\ x_n[k+1] \end{bmatrix} \tag{5.159}
$$

Substituting the state vector in the left side of 5.158 by 5.159 we get

$$
\begin{bmatrix} x_1[k+2] \\ \vdots \\ x_n[k+2] \end{bmatrix} = \Phi^2 \begin{bmatrix} x_1[k] \\ \vdots \\ x_n[k] \end{bmatrix} \tag{5.160}
$$

We can go on delaying and get

$$
\begin{bmatrix} x_1[k+n-1] \\ \vdots \\ x_n[k+n-1] \end{bmatrix} = \Phi^{n-1} \begin{bmatrix} x_1[k] \\ \vdots \\ x_n[k] \end{bmatrix} \tag{5.161}
$$

If we observe the output given by

$$\mathbf{y}[k] = \mathbf{C} \begin{bmatrix} x_1[k] \\ \vdots \\ x_n[k] \end{bmatrix} \tag{5.162}$$

and measure it delayed, we get

$$\mathbf{y}[k+1] = \mathbf{C} \begin{bmatrix} x_1[k+1] \\ \vdots \\ x_n[k+1] \end{bmatrix} = \mathbf{C\Phi} \begin{bmatrix} x_1[k] \\ \vdots \\ x_n[k] \end{bmatrix}$$

$$\mathbf{y}[k+2] = \mathbf{C} \begin{bmatrix} x_1[k+2] \\ \vdots \\ x_n[k+2] \end{bmatrix} = \mathbf{C\Phi}^2 \begin{bmatrix} x_1[k] \\ \vdots \\ x_n[k] \end{bmatrix}$$

$$\vdots$$

$$\mathbf{y}[k+n-1] = \mathbf{C} \begin{bmatrix} x_1[k+n-1] \\ \vdots \\ x_n[k+n-1] \end{bmatrix} = \mathbf{C\Phi}^{n-1} \begin{bmatrix} x_1[k] \\ \vdots \\ x_n[k] \end{bmatrix} \tag{5.163}$$

By the definition of observability, given the n measures of outputs, we must be able to find the state at any instant k. In other words, we must be able to solve the following system (result of writing 5.163 in matrix form)

$$\begin{bmatrix} \mathbf{y}[k] \\ \mathbf{y}[k+1] \\ \vdots \\ \mathbf{y}[k+n-1] \end{bmatrix} = \begin{bmatrix} \mathbf{C} \\ \mathbf{C\Phi} \\ \vdots \\ \mathbf{C\Phi}^{n-1} \end{bmatrix} \begin{bmatrix} x_1[k] \\ \vdots \\ x_n[k] \end{bmatrix} \tag{5.164}$$

Defining

$$\mathbf{O} = \begin{bmatrix} \mathbf{C} \\ \mathbf{C\Phi} \\ \vdots \\ \mathbf{C\Phi}^{n-1} \end{bmatrix} \tag{5.165}$$

we conclude that, in order to have a solution, $rank(\mathbf{O}) = n$
∎.

5.7.2   Controllability

Continuous systems

To analyse the controllability of a given representation, we
must apply a particular input signal at a given instant t and
be able to go from any state $\mathbf{x}_i$ to another state $\mathbf{x}_f$. In other
words, we have to make sure that the system can be "guided"
to a desired final state by means of applying an input at any
instant t. By "particular input signal" we mean a signal that we
choose having any value as well as derivative value.

Following the same idea as in the observability test, we start
the state space representation given by

$$\dot{\mathbf{x}}(t) = \mathbf{A}\mathbf{x}(t) + \mathbf{B}\mathbf{u}(t)$$

$$\begin{bmatrix} \dot{x}_1(t) \\ \vdots \\ \dot{x}_n(t) \end{bmatrix} = \mathbf{A} \begin{bmatrix} x_1(t) \\ \vdots \\ x_n(t) \end{bmatrix} + \mathbf{B}\mathbf{u}(t) \tag{5.166}$$

Differentiating both sides we have

$$\begin{bmatrix} \ddot{x}_1(t) \\ \vdots \\ \ddot{x}_n(t) \end{bmatrix} = \mathbf{A} \begin{bmatrix} \dot{x}_1(t) \\ \vdots \\ \dot{x}_n(t) \end{bmatrix} + \mathbf{B}\dot{\mathbf{u}}(t) \tag{5.167}$$

Substituting 5.166 into 5.167 we have

$$\begin{bmatrix} \ddot{x}_1(t) \\ \vdots \\ \ddot{x}_n(t) \end{bmatrix} = \mathbf{A} \left(\mathbf{A} \begin{bmatrix} x_1(t) \\ \vdots \\ x_n(t) \end{bmatrix} + \mathbf{B}\mathbf{u}(t) \right) + \mathbf{B}\dot{\mathbf{u}}(t)$$

$$= \mathbf{A}^2 \begin{bmatrix} x_1(t) \\ \vdots \\ x_n(t) \end{bmatrix} + \mathbf{A}\mathbf{B}\mathbf{u}(t) + \mathbf{B}\dot{\mathbf{u}}(t) \tag{5.168}$$

If we keep differentiating, we can obtain

$$\begin{bmatrix} x_1^{(n)}(t) \\ \vdots \\ x_n^{(n)}(t) \end{bmatrix} = \mathbf{A}^n \begin{bmatrix} x_1(t) \\ \vdots \\ x_n(t) \end{bmatrix} + \mathbf{A}^{n-1}\mathbf{B}u(t) + \ldots + \mathbf{B}u^{n-1}(t)$$

(5.169)

Reordering the terms, we obtain

$$\begin{bmatrix} \mathbf{B} & \mathbf{AB} & \cdots & \mathbf{A}^{n-1}\mathbf{B} \end{bmatrix} \begin{bmatrix} u^{(n-1)}(t) \\ u^{(n-2)}(t) \\ \vdots \\ u(t) \end{bmatrix} = \begin{bmatrix} x_1^{(n)}(t) \\ x_2^{(n)}(t) \\ \vdots \\ x_n^{(n)}(t) \end{bmatrix} - \mathbf{A}^n \begin{bmatrix} x_1(t) \\ x_2(t) \\ \vdots \\ x_n(t) \end{bmatrix}$$

(5.170)

and we can define $\mathbf{R} = \begin{bmatrix} \mathbf{B} & \mathbf{AB} & \cdots & \mathbf{A}^{n-1}\mathbf{B} \end{bmatrix}$.

Following the concept of controllability, in order for a representation to be controllable, we must be able to find the input signal ($\mathbf{u}(t)$ and its derivatives) that "guide" the state to any point (having a state $\mathbf{x}(t)$ and its derivatives). To do that, the system of equations presented in 5.170 must have a solution and therefore, its rank must be n ■.

Discrete time systems

Following the continuous case, consider the state space now represented by (we now need the input because of the definition)

$$\begin{bmatrix} x_1[k+1] \\ \vdots \\ x_n[k+1] \end{bmatrix} = \Phi \begin{bmatrix} x_1[k] \\ \vdots \\ x_n[k] \end{bmatrix} + \Gamma\mathbf{u}[k]$$

(5.171)

One delay ahead, we have

$$\begin{bmatrix} x_1[k+2] \\ \vdots \\ x_n[k+2] \end{bmatrix} = \Phi \begin{bmatrix} x_1[k+1] \\ \vdots \\ x_n[k+1] \end{bmatrix} + \Gamma \mathbf{u}[k+1] \qquad (5.172)$$

Substituting the right side of 5.171 into 5.172 and repeating the process, we obtain (analogous to the observability case)

$$\begin{bmatrix} x_1[k+n-1] \\ \vdots \\ x_n[k+n-1] \end{bmatrix} = \Phi^{n-1} \begin{bmatrix} x_1[k] \\ \vdots \\ x_n[k] \end{bmatrix} +$$

$$\Phi^{n-2}\Gamma \mathbf{u}[k] + \Phi^{n-3}\Gamma \mathbf{u}[k+1] + \ldots + \Gamma \mathbf{u}[k+n-1] \quad (5.173)$$

Rewriting, we get

$$\begin{bmatrix} \Gamma & \Phi\Gamma & \ldots & \Phi^{n-1}\Gamma \end{bmatrix} \begin{bmatrix} u[k+n-1] \\ u[k+n-2] \\ \vdots \\ u[k] \end{bmatrix} =$$

$$\begin{bmatrix} x_1[k+n-1] \\ x_2[k+n-1] \\ \vdots \\ x_n[k+n-1] \end{bmatrix} - \Phi^{n-1} \begin{bmatrix} x_1[k] \\ x_2[k] \\ \vdots \\ x_n[k] \end{bmatrix} \qquad (5.174)$$

so that we can define

$$\mathbf{R} = \begin{bmatrix} \Gamma & \Phi\Gamma & \ldots & \Phi^{n-1}\Gamma \end{bmatrix} \qquad (5.175)$$

Therefore, for the representation to be controllable, we must be able to go from a state to another after n time steps. That means that the system 5.174 must be solvable, and that implies $rank(\mathbf{R}) = n$.

5.8 State observer

In this section we will derive the equations to analyse and design state observers for a certain dynamical system state space representation. We will show the equations for the design of both continuous and discrete systems.

5.8.1 Continuous case

We start with the state space representation for a LTI system

$$\dot{\mathbf{x}}(t) = \mathbf{A}\mathbf{x}(t) + \mathbf{B}u(t)$$
$$y(t) = \mathbf{C}\mathbf{x}(t) + \mathbf{D}u(t) \tag{5.176}$$

The idea behind the observer is that a set of estimated states $\hat{\mathbf{x}}(t)$ will follow the states $\mathbf{x}(t)$ as t progresses. To correct the estimated state, a correction term depending on the difference between the outputs is added to its dynamical equation. The equations for the dynamic of estimated states is then

$$\dot{\hat{\mathbf{x}}}(t) = \mathbf{A}\hat{\mathbf{x}}(t) + \mathbf{B}u(t) + \mathbf{L}(y(t) - \hat{y}(t))$$
$$\hat{y}(t) = \mathbf{C}\hat{\mathbf{x}}(t) + \mathbf{D}u(t) \tag{5.177}$$

The gain vector $\mathbf{L}$ is chosen to ensure that the estimated state will converge to the systemâĂŹs actual state.

In order to project the observer gain vector, we will substitute the output for the real system into the estimator. That leads us to

$$\dot{\hat{\mathbf{x}}}(t) = \mathbf{A}\hat{\mathbf{x}}(t) + \mathbf{B}u(t) + \mathbf{L}(\mathbf{C}\mathbf{x}(t) + \mathbf{D}u(t) - \mathbf{C}\hat{\mathbf{x}}(t) - \mathbf{D}u(t))$$
$$\dot{\hat{\mathbf{x}}}(t) = \mathbf{A}\hat{\mathbf{x}}(t) + \mathbf{B}u(t) + (\mathbf{L}\mathbf{C}(\mathbf{x}(t) - \hat{\mathbf{x}}(t))) \tag{5.178}$$

Subtracting $\dot{\mathbf{x}}(t)$ from both sides (but using the transition equation in the right side), we have

$$\dot{\hat{\mathbf{x}}}(t) - \dot{\mathbf{x}}(t) = \mathbf{A}\hat{\mathbf{x}}(t) + \mathbf{B}u(t) + (\mathbf{L}\mathbf{C}(\mathbf{x}(t) - \hat{\mathbf{x}}(t))) - \mathbf{A}\mathbf{x}(t) - \mathbf{B}u(t) \tag{5.179}$$

which develops to

$$
\begin{aligned}
\dot{\hat{\mathbf{x}}}(t) - \dot{\mathbf{x}}(t) &= -\mathbf{A}(\mathbf{x}(t) - \hat{\mathbf{x}}(t)) + (\mathbf{LC}(\mathbf{x}(t) - \hat{\mathbf{x}}(t)) \\
\dot{\hat{\mathbf{x}}}(t) - \dot{\mathbf{x}}(t) &= (\mathbf{A} - \mathbf{LC})(\hat{\mathbf{x}}(t) - \mathbf{x}(t))
\end{aligned}
\tag{5.180}
$$

Now, we define the error state as

$$
\mathbf{e}(t) = \hat{\mathbf{x}}(t) - \mathbf{x}(t)
\tag{5.181}
$$

and rewrite the dynamic equation for the state difference as

$$
\dot{\mathbf{e}}(t) = (\mathbf{A} - \mathbf{LC})\mathbf{e}(t)
\tag{5.182}
$$

Now, we have a dynamical evolution for the state's error. We can choose $\mathbf{L}$ such that this evolution is stable and fast. To do that we force the eigenvalues of the matrix $\mathbf{A} - \mathbf{L}\,\mathbf{C}$ to be some numbers with a negative real part. Thus, to project the observer gains $\mathbf{L}$ we need to solve the following set of equations

$$
\det(\lambda_i \mathbf{I} - \mathbf{A} + \mathbf{LC}) = 0, \quad i = 1, \ldots, n
\tag{5.183}
$$

where n is the order of the system (size of $\mathbf{A}$) and the λ_is are the observer poles (chosen to have negative real parts).

5.8.2   Discrete case

As in the continuous case, we start with the state space representation

$$
\begin{aligned}
\mathbf{x}[k + 1] &= \Phi\mathbf{x}[k] + \Gamma u[k] \\
y[k] &= \mathbf{C}\mathbf{x}[k] + Du[k]
\end{aligned}
\tag{5.184}
$$

Also as in the continuous case, we wish to implement a structure that tracks the state's values by measuring only the output of the system. The structure (similar to the continuous case) is given by

$$\hat{\mathbf{x}}[k + 1] = \Phi\hat{\mathbf{x}}[k] + \Gamma u[k] + \mathbf{L}\left(y[k] - \hat{y}[k]\right)$$
$$\hat{y}[k] = C\hat{\mathbf{x}}[k] + Du[k] \tag{5.185}$$

Now, we plug in the expressions for the real and estimated output as

$$\hat{\mathbf{x}}[k+1] = \Phi\hat{\mathbf{x}}[k] + \Gamma u[k] + \mathbf{L}\left((\mathbf{C}\mathbf{x}[k] + Du[k]) - (C\hat{\mathbf{x}}[k] + Du[k])\right) \tag{5.186}$$

which leads to

$$\hat{\mathbf{x}}[k + 1] = \Phi\hat{\mathbf{x}}[k] + \Gamma u[k] + \mathbf{L}\mathbf{C}\left(\mathbf{x}[k] - \hat{\mathbf{x}}[k]\right) \tag{5.187}$$

Now, we subtract $\mathbf{x}[k + 1]$ from both sides

$$\hat{\mathbf{x}}[k+1] - \mathbf{x}[k+1] = \Phi\hat{\mathbf{x}}[k] + \Gamma u[k] + \mathbf{L}\mathbf{C}\left(\mathbf{x}[k] - \hat{\mathbf{x}}[k]\right) - (\Phi\mathbf{x}[k] + \Gamma u[k]) \tag{5.188}$$

and have

$$\hat{\mathbf{x}}[k + 1] - \mathbf{x}[k + 1] = \Phi\left(\hat{\mathbf{x}}[k] - \mathbf{x}[k]\right) - \mathbf{L}\mathbf{C}\left(\hat{\mathbf{x}}[k] - \mathbf{x}[k]\right) \tag{5.189}$$

Now, we define the state error as

$$\mathbf{e}[k] = \hat{\mathbf{x}}[k] - \mathbf{x}[k] \tag{5.190}$$

Now, we plug in the definition into 5.189 and have

$$\mathbf{e}[k + 1] = (\Phi - \mathbf{L}\mathbf{C})\,\mathbf{e}[k] \tag{5.191}$$

that has the following solution

$$\mathbf{e}[k] = \mathbf{e}_0(\Phi - \mathbf{L}\mathbf{C})^k \tag{5.192}$$

Writing

$$\Phi - \mathbf{L}\mathbf{C} = \mathbf{M} \tag{5.193}$$

we can have the following eigen-decomposition

$$\mathbf{M}^k = \mathbf{U}\mathbf{D}^k\mathbf{U}^t \tag{5.194}$$

with

$$\mathbf{D}^k = \begin{bmatrix} \lambda_1^k & 0 & 0 \\ 0 & \ddots & 0 \\ 0 & 0 & \lambda_N^k \end{bmatrix} \tag{5.195}$$

λ_i being the eigenvalues of $\mathbf{M} = \Phi - \mathbf{L}\mathbf{C}$. So to design an observer we pick values for the eigenvalues satisfying

$$|\lambda_i| < 1, \quad i = 1, ..., N \tag{5.196}$$

and solve

$$\det(\lambda\mathbf{I} - \Phi + \mathbf{L}\mathbf{C}) = 0 \tag{5.197}$$

for $\mathbf{L}$ ∎.

5.9   Transfer function from state space

5.9.1   Transfer function

From a state space representation

$$\begin{aligned} \dot{\mathbf{x}}(t) &= \mathbf{A}\mathbf{x}(t) + \mathbf{B}u(t) \\ y(t) &= \mathbf{C}\mathbf{x}(t) + Du(t) \end{aligned} \tag{5.198}$$

we can obtain the transfer function by using the Laplace transform in the state equations (and considering zero initial conditions). We will obtain

$$\begin{aligned} s\mathbf{x}(s) &= \mathbf{A}\mathbf{x}(s) + \mathbf{B}U(s) \\ Y(s) &= \mathbf{C}\mathbf{x}(s) + DU(s) \end{aligned} \tag{5.199}$$

Now, we can obtain $\mathbf{x}(s)$ from the transition equation as

$$
\begin{aligned}
s\mathbf{x}(s) - \mathbf{A}\mathbf{x}(s) &= \mathbf{B}U(s) \\
(s\mathbf{I} - \mathbf{A})\mathbf{x}(s) &= \mathbf{B}U(s) \\
\mathbf{x}(s) &= (s\mathbf{I} - \mathbf{A})^{-1}\mathbf{B}U(s)
\end{aligned}
\tag{5.200}
$$

We can plug 5.200 into the output equation and finally get

$$
Y(s) = \mathbf{C}((s\mathbf{I} - \mathbf{A})^{-1}\mathbf{B}U(s)) + DU(s) \tag{5.201}
$$

that leads to the transfer function ■

$$
\frac{Y(s)}{U(s)} = \mathbf{C}(s\mathbf{I} - \mathbf{A})^{-1}\mathbf{B} + D \tag{5.202}
$$

Invariance under linear transformation

If we have some state space representation for a given system, the states could represent some physical variable inside the system. In general, the states could represent not only physical quantities but combinations of internal variables like currents, positions, etc. One could "transform" the states into a new set of values with an invertible linear transformation as

$$
\hat{\mathbf{x}}(t) = \mathbf{P}\mathbf{x}(t) \tag{5.203}
$$

Now, if we write the new state space representation, we have

$$
\begin{aligned}
\mathbf{P}^{-1}\dot{\hat{\mathbf{x}}}(t) &= \mathbf{A}\mathbf{P}^{-1}\hat{\mathbf{x}}(t) + \mathbf{B}u(t) \\
y(t) &= \mathbf{C}\mathbf{P}^{-1}\hat{\mathbf{x}}(t) + Du(t)
\end{aligned}
\tag{5.204}
$$

which translates to

$$
\begin{aligned}
\dot{\hat{\mathbf{x}}}(t) &= \mathbf{P}\mathbf{A}\mathbf{P}^{-1}\hat{\mathbf{x}}(t) + \mathbf{P}\mathbf{B}u(t) \\
y(t) &= \mathbf{C}\mathbf{P}^{-1}\hat{\mathbf{x}}(t) + Du(t)
\end{aligned}
\tag{5.205}
$$

If we use 5.202 to compute the new transfer function, we will have

$$\frac{\hat{Y}(s)}{\hat{U}(s)} = \mathbf{CP}^{-1}(s\mathbf{I} - \mathbf{PAP}^{-1})^{-1}\mathbf{PB} + D \qquad (5.206)$$

But, we can write the identity as the product of the transformation and its inverse as

$$\frac{\hat{Y}(s)}{\hat{U}(s)} = \mathbf{CP}^{-1}(s\mathbf{PP}^{-1} - \mathbf{PAP}^{-1})^{-1}\mathbf{PB} + D \qquad (5.207)$$

which develops to ■

$$\frac{\hat{Y}(s)}{\hat{U}(s)} = \mathbf{CP}^{-1}\mathbf{P}(s\mathbf{I} - \mathbf{A})^{-1}\mathbf{P}^{-1}\mathbf{PB} + D$$
$$= \mathbf{C}(s\mathbf{I} - \mathbf{A})^{-1}\mathbf{B} + D = \frac{Y(s)}{U(s)} \qquad (5.208)$$

That result tells us that the transfer function is invariant to a linear transformation in the states.

Chapter 6

Machine Learning

Evolution of computer power and capacity has facilitated the creation and development of many new branches of science and technology. One of these branches is machine learning.

6.1 Learning Vector Quantisation

The learning vector quantisation (LVQ) algorithm is used in traditional clustering methods and is at the heart of competitive neural networks. When one treats the LVQ as an optimisation problem, one of the costs or functionals that is frequently used is the intra-cluster distance given by

$$E = \sum_k \|\mathbf{x}_k - \mathbf{m}_c\|^2 \tag{6.1}$$

where $\mathbf{x}_k$ are the data points and $\mathbf{m}_i$ is the j-th centre. $\mathbf{m}_c$ is the centre that is closest to $\mathbf{x}_k$, which gives

$$\|\mathbf{x}_k - \mathbf{m}_c\|^2 = \min_i \|\mathbf{x}_k - \mathbf{m}_i\|^2 \tag{6.2}$$

The goal is to find the vector $\mathbf{m}_c$ that minimises that cost. But this is a discontinuous function and therefore it is hard to minimise with respect to this variable. Utilising the zero norm of a sequence given by

$$\min_i a_i = \lim_{r \to -\infty} \left(\sum_i a_i^r \right)^{\frac{1}{r}} \tag{6.3}$$

we can consider the sequence of i distances $\|\mathbf{x}_k - \mathbf{m}_i\|^2$. We then have

$$\|\mathbf{x}_k - \mathbf{m}_c\|^2 = \min_i \|\mathbf{x}_k - \mathbf{m}_i\|^2 = \lim_{r \to -\infty} \left(\sum_i \|\mathbf{x}_k - \mathbf{m}_i\|^{2r} \right)^{\frac{1}{r}} \tag{6.4}$$

In this manner we can write the cost function as

$$E = \lim_{r \to -\infty} \sum_k \left(\sum_i \|\mathbf{x}_k - \mathbf{m}_i\|^{2r} \right)^{\frac{1}{r}} \tag{6.5}$$

Now, except in the limit, this is a continuous function and can be differentiated wrt $\mathbf{m}_i$. Let us take the gradient wrt one of the centres $\mathbf{m}_j$. We get

$$\nabla_{m_j} E = \lim_{r \to -\infty} \sum_k \nabla_{m_j} \left(\sum_i \|\mathbf{x}_k - \mathbf{m}_i\|^{2r} \right)^{\frac{1}{r}} \tag{6.6}$$

Computing this gradient we have

$$\nabla_{m_j} E = \lim_{r \to -\infty} \sum_k \frac{1}{r} \left(\sum_i \|\mathbf{x}_k - \mathbf{m}_i\|^{2r} \right)^{\frac{1}{r}-1} \nabla_{m_j} \sum_i \|\mathbf{x}_k - \mathbf{m}_i\|^{2r} \tag{6.7}$$

Using the chain rule we get

$$\nabla_{m_j} E = \lim_{r \to -\infty} \sum_k \frac{1}{r} \left(\sum_i \|\mathbf{x}_k - \mathbf{m}_i\|^{2r} \right)^{\frac{1}{r}-1} \nabla_{m_j} \|\mathbf{x}_k - \mathbf{m}_j\|^{2r} \tag{6.8}$$

Developing a little we obtain

$$\nabla_{m_j} E = \lim_{r \to -\infty} \sum_k \frac{1}{r} \left(\sum_i \|\mathbf{x}_k - \mathbf{m}_i\|^{2r} \right)^{\frac{1}{r}-1} 2r \|\mathbf{x}_k - \mathbf{m}_j\|^{2r-1} \nabla_{m_j} \|\mathbf{x}_k - \mathbf{m}_j\|$$

(6.9)

which gives us

$$\nabla_{m_j} E = -2 \lim_{r \to -\infty} \sum_k \left(\sum_i \|\mathbf{x}_k - \mathbf{m}_i\|^{2r} \right)^{\frac{1}{r}-1} \|\mathbf{x}_k - \mathbf{m}_j\|^{2r-2} (\mathbf{x}_k - \mathbf{m}_j)$$

(6.10)

Now we proceed to write the cost function in a way that is easier to manipulate (by generating a summation term that will be useful later) as

$$\nabla_{m_j} E =$$

$$-2 \lim_{r \to -\infty} \sum_k \left(\sum_i \|\mathbf{x}_k - \mathbf{m}_i\|^{2r} \right)^{\frac{1}{r}-1} \left(\|\mathbf{x}_k - \mathbf{m}_j\|^{(2r-2)\left(\frac{1}{\frac{1}{r}-1}\right)} \right)^{\frac{1}{r}-1} (\mathbf{x}_k - \mathbf{m}_j)$$

(6.11)

Simplifying gives

$$\nabla_{m_j} E = -2 \lim_{r \to -\infty} \sum_k \left(\sum_i \|\mathbf{x}_k - \mathbf{m}_i\|^{2r} \|\mathbf{x}_k - \mathbf{m}_j\|^{-2r} \right)^{\frac{1}{r}-1} (\mathbf{x}_k - \mathbf{m}_j)$$

(6.12)

and finally

$$\nabla_{m_j} E = -2 \sum_k (\mathbf{x}_k - \mathbf{m}_j) \lim_{r \to -\infty} \left(\sum_i \frac{\|\mathbf{x}_k - \mathbf{m}_i\|^{2r}}{\|\mathbf{x}_k - \mathbf{m}_j\|^{2r}} \right)^{\frac{1}{r}-1}$$

(6.13)

The term in this last form of the cost function is very interesting. This term depends only on $\mathbf{x}_k$ and $\mathbf{m}_j$. We can verify that if $\mathbf{x}_k$ is closest to $\mathbf{m}_j$, then there will always exist a denominator less than one, and that will make the term (in the limit $r \to -\infty$) tend to zero. By inspection, this term is a Kronecker delta in k and j. So the gradient becomes

$$\nabla_{m_j} E = -2 \sum_k (\mathbf{x}_k - \mathbf{m}_j)\, \delta_{j,k} \qquad (6.14)$$

As the delta has a unitary value when $j = k$, we write

$$\nabla_{m_j} E = -2 (\mathbf{x}_k - \mathbf{m}_c) \qquad (6.15)$$

Here, c is the index of the point that is closest to $\mathbf{x}_k$.

Now using the gradient descent algorithm to update the values of each $\mathbf{m}_j$, we are left with the following centre update:

$$\mathbf{m}_j[k+1] = \mathbf{m}_j[k] + 2\alpha (\mathbf{x}_k - \mathbf{m}_j[k]) \qquad (6.16)$$

in which each iteration each $\mathbf{m}_j$ is updated in the direction of the point $\mathbf{x}_k$ closest to it.

It is important to notice that this algorithm will take the centres to a local minimum on $\mathbf{x}_k$. To avoid this, at each iteration $\mathbf{x}_k$ is exchanged by another point in the data set and the closest centre is updated. This leads to the classical LVQ algorithm.

6.1.1 Limit Analysis

The expression

$$\lim_{r \to -\infty} \left(\sum_i \frac{\|\mathbf{x}_k - \mathbf{m}_i\|^{2r}}{\|\mathbf{x}_k - \mathbf{m}_j\|^{2r}} \right)^{\frac{1}{r}-1} \qquad (6.17)$$

can be rewritten (using the fact that $\lim_{r \to -\infty} \frac{1}{r} = 0$) as

$$\lim_{r \to -\infty} \left(\sum_i \frac{\|\mathbf{x}_k - \mathbf{m}_i\|^{2r}}{\|\mathbf{x}_k - \mathbf{m}_j\|^{2r}} \right)^{\frac{1}{r} - 1} =$$

$$\lim_{r \to -\infty} \left(\sum_i \left(\frac{\|\mathbf{x}_k - \mathbf{m}_i\|^2}{\|\mathbf{x}_k - \mathbf{m}_j\|^2} \right)^r \right)^{-1} = \frac{1}{\sum_i \lim_{r \to \infty} \left(\frac{\|\mathbf{x}_k - \mathbf{m}_j\|^2}{\|\mathbf{x}_k - \mathbf{m}_i\|^2} \right)^r}$$

$$(6.18)$$

Thus, it is equal to one only for the $\mathbf{x}_k$ that is closest to $\mathbf{m}_j$. Next we will provide an example.

Example

Consider the situation illustrated in figure 6.1 with four centres.

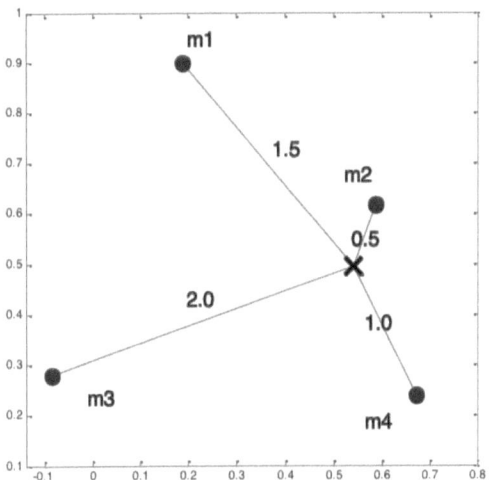

Figure 6.1: Several centres (circles) around a point $\mathbf{x}_k$ ("x" in the figure). Distances are shown for each centre.

The limit in question can be computed as

$$\frac{1}{A_1 + A_2 + A_3 + A_4}$$

where

$$A_1 = \lim_{r \to \infty} \left(\frac{\|\mathbf{x}_k - \mathbf{m}_j\|^2}{\|\mathbf{x}_k - \mathbf{m}_1\|^2} \right)^r$$

$$A_2 = \lim_{r \to \infty} \left(\frac{\|\mathbf{x}_k - \mathbf{m}_j\|^2}{\|\mathbf{x}_k - \mathbf{m}_2\|^2} \right)^r$$

$$A_3 = \lim_{r \to \infty} \left(\frac{\|\mathbf{x}_k - \mathbf{m}_j\|^2}{\|\mathbf{x}_k - \mathbf{m}_3\|^2} \right)^r \tag{6.19}$$

$$A_4 = \lim_{r \to \infty} \left(\frac{\|\mathbf{x}_k - \mathbf{m}_j\|^2}{\|\mathbf{x}_k - \mathbf{m}_4\|^2} \right)^r$$

For $j = 4$ we have

$$\frac{1}{\lim_{r \to \infty} \left(\frac{1.0}{1.5} \right)^r + \lim_{r \to \infty} \left(\frac{1.0}{0.5} \right)^r + \lim_{r \to \infty} \left(\frac{1.0}{2.0} \right)^r + \lim_{r \to \infty} \left(\frac{1.0}{1.0} \right)^r} \tag{6.20}$$

$$= \frac{1}{0 + \infty + 0 + 1} = 0$$

For $j = 3$ we have

$$\frac{1}{\lim_{r \to \infty} \left(\frac{0.5}{1.5} \right)^r + \lim_{r \to \infty} \left(\frac{0.5}{0.5} \right)^r + \lim_{r \to \infty} \left(\frac{0.5}{2.0} \right)^r + \lim_{r \to \infty} \left(\frac{0.5}{1.0} \right)^r} \tag{6.21}$$

$$= \frac{1}{0 + 1 + 0 + 0} = 1$$

For $j = 2$ we have

$$\frac{1}{\lim_{r \to \infty} \left(\frac{0.5}{1.5} \right)^r + \lim_{r \to \infty} \left(\frac{0.5}{0.5} \right)^r + \lim_{r \to \infty} \left(\frac{0.5}{2.0} \right)^r + \lim_{r \to \infty} \left(\frac{0.5}{1.0} \right)^r} \tag{6.22}$$

$$= \frac{1}{0 + 1 + 0 + 0} = 1$$

For $j = 1$ we have

$$\frac{1}{\lim_{r\to\infty}\left(\frac{1.5}{1.5}\right)^r + \lim_{r\to\infty}\left(\frac{1.5}{0.5}\right)^r + \lim_{r\to\infty}\left(\frac{1.5}{2.0}\right)^r + \lim_{r\to\infty}\left(\frac{1.5}{1.0}\right)^r} \tag{6.23}$$

$$= \frac{1}{1 + \infty + 0 + \infty} = 0$$

As we can see, only when $\mathbf{m}_j$ is the centre closest to $\mathbf{x}_k$ ($j = 2$) do we have a value that is unitary, otherwise we get zero.

6.2 Principal Component Analysis

Principal component analysis (PCA) is a linear technique that finds linear projections of a data set $X = \{\mathbf{x}_1, \mathbf{x}_2, \ldots, \mathbf{x}_N\}$ ($\mathbf{x}_i \in \Re^n$) so that those projections have maximised (and sorted) variances for each projected variable. The requirement is that the expected value of the data set is zero ($E\{X\} = 0$) and we maximise the variance of the projected data keeping the norm finite. We must them find a new set of projected values $y_i = \mathbf{w}^t \mathbf{x}_i$ such that

$$\mathbf{w} = \arg\max_{\|\mathbf{w}\|^2 = 1} \text{Var}\{\mathbf{w}^t \mathbf{X}\} \tag{6.24}$$

To do so, we maximise the functional

$$\begin{aligned}
\varepsilon &= \text{Var}\{\mathbf{w}^t \mathbf{X}\} + \lambda\left(\|\mathbf{w}\|^2 - 1\right) \\
&= \frac{1}{N}\sum_{i=1}^{N}\left(\mathbf{w}^t \mathbf{x}_i\right)^2 + \lambda\left(\|\mathbf{w}\|^2 - 1\right)
\end{aligned} \tag{6.25}$$

Differentiating we have

$$\frac{\partial\varepsilon}{\partial\mathbf{w}} = \frac{2}{N}\sum_{i=1}^{N}\mathbf{w}^t\mathbf{x}_i\mathbf{x}_i^t + 2\lambda\mathbf{w}^t \tag{6.26}$$

$$\frac{\partial\varepsilon}{\partial\lambda} = \|\mathbf{w}\|^2 - 1$$

The second equation is the norm restriction itself. Equating the first equation to zero we obtain

$$\mathbf{w}^t \frac{1}{N} \sum_{i=1}^{N} \mathbf{x}_i \mathbf{x}_i^t = -\lambda \mathbf{w}^t \qquad (6.27)$$

The term of the summation on the right side of the equation is the covariance matrix of the data X (Σ_X). We can now write it in the following compact form:

$$\mathbf{w}^t \Sigma_X = \lambda \, \mathbf{w}^t \qquad (6.28)$$

This equation defines the eigenvalues and eigenvectors of the matrix Σ_X. It is clear that the vector that satisfies 6.24 will be the eigenvector corresponding to the largest eigenvalue of the covariance matrix. We pick the largest because from the beginning, we wish to maximise the functional. Since all the vectors are solutions, we must pick the largest one.

From the result, we notice that there are n solutions to equation 6.28. The solutions are ordered by the magnitude of the eigenvalues of the covariance matrix and thus we get a set of projections ordered by the variances.

6.3   Regularisation Theory

One of the big problems for any learning machine is overfitting. If one minimises the error in some regression problem to the point that the function passes exactly for all points, there is a big chance that the function will not behave well for points where the function was not trained. Figure 6.2 illustrates an example. A better regression or approximation would be a function that instead of passing through all the points passes with the smallest error possible while keeping some degree of smoothness (figure 6.2(b)). The goal of a regularised RBF is to minimise the error, while keeping some degree of smoothness for the function. To do so, we have to have some measure of how smooth the function is. This is achieved by measuring the norm of some differential or derivative of the function (considering it as a vector in some Hilbert space $H = f : R \rightarrow \Re$).

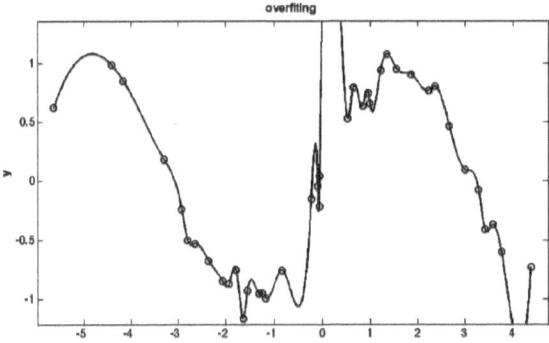

(a) Example of overfitted data. Points are the available data and the line is the fitted function.

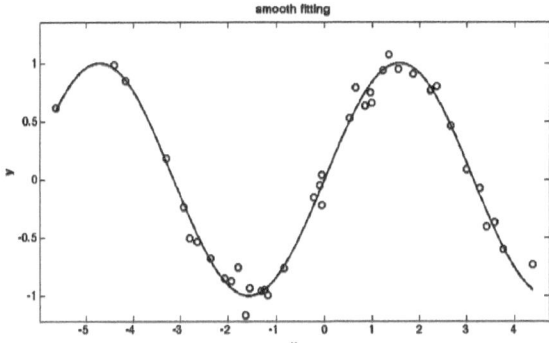

(b) Example of correct fitting. Points are the available data and the line is the fitted function.

Figure 6.2: Examples of function fit. a) Overfitted function and b) correct fitting.

We are interested in minimising the functional

$$J = \sum_i (f(x_i) - y_i)^2 + \lambda \|Df(x)\|^2 \qquad (6.29)$$

Here, $f(x)$ is the function we wish to find, and (x_i, y_i) are the pairs of known data we wish to fit $f(x)$ onto. The operator D is a differential operator and $\|.\|^2$ is the Hilbert space square norm of the argument function. λ is the weight factor that will favour smoothness against the fitting error.

From functional analysis, we want to displace $f(x)$ by a very small amount $e(x)$ so that this displacement will cause a displacement of ΔJ onto the functional. We them have

$$J+\Delta J = \sum_i (f(x_i) + e(x_i) - y_i)^2 + \lambda \|Df(x) + De(x)\|^2 \quad (6.30)$$

Now we expand the terms and apply the integral for the square norm (in the whole domain R of $f(x)$) of the second term to get

$$J + \Delta J = \sum_i (f(x_i) - y_i)^2 + 2\sum_i (f(x_i) - y_i)\, e(x_i) + e(x_i)^2 +$$
$$\lambda \int_R (Df(x))^2 + 2Df(x)De(x) + (De(x))^2 \mathrm{d}x$$

$$(6.31)$$

The next step is to make the displacement function as small as we can, which makes some of the displacement terms go to zero. We then get

$$J + \Delta J = \sum_i (f(x_i) - y_i)^2 +$$
$$\lambda \int_R (Df(x))^2 \mathrm{d}x + 2\sum_i (f(x_i) - y_i)\, e(x_i) + \lambda \int_R 2Df(x)De(x)\mathrm{d}x$$

$$(6.32)$$

Now we obtain ΔJ by identifying the right-hand side, leaving us with:

$$\Delta J = 2\sum_i (f(x_i) - y_i)\, e(x_i) + \lambda \int_R 2Df(x)De(x)\mathrm{d}x \quad (6.33)$$

Now, as $e(x)$ goes to zero, we need to make sure that ΔJ also goes to zero so that we are in a flat region of the Hilbert space. This leads to:

$$0 = \sum_i \left(f(x_i) - y_i\right) e(x_i) + \lambda \int_R Df(x)De(x)\mathrm{d}x \qquad (6.34)$$

The problem is that the term in the integral might not go to zero (because it is an integral of some small quantity), and so we have to make sure we have one integral with an integrand of zero. The trick here is to write the evaluated quantity $e(x_i)$ as an integral. We do so using a delta function as

$$0 = \sum_i \left(f(x_i) - y_i\right) \int_R e(x)\delta(x - x_i)\mathrm{d}x + \lambda \int_R Df(x)De(x)\mathrm{d}x$$
$$(6.35)$$

Another problem is that the quantity $De(x)$ does not goes to zero in the same fashion that the quantity $e(x)$ does. This makes the solution for $f(x)$ dependent on $e(x)$ (or $De(x)$), which is not acceptable. To circumvent this difficulty, we write the second integral using integration by parts. To do so, we choose the functions:

$$\begin{aligned} v &= Df(x) \\ \mathrm{d}u &= De(x)\mathrm{d}x \end{aligned} \qquad (6.36)$$

Changing the integrands we have (we integrate and differentiate in the domain R of $f(x)$)

$$\begin{aligned} \mathrm{d}v &= D^2f(x)\mathrm{d}x \\ u &= e(x) \end{aligned} \qquad (6.37)$$

Now we are left with:

$$\int_R Df(x)De(x)\mathrm{d}x = e(x)Df(x) - \int_R D^2f(x)e(x)\mathrm{d}x \qquad (6.38)$$

Going back to the original equation we have

$$\sum_i (f(x_i) - y_i) \int_R e(x)\delta(x - x_i)\mathrm{d}x + \lambda e(x)\mathrm{D}f(x) - \lambda \int_R \mathrm{D}^2 f(x)e(x)\mathrm{d}x$$

$$= \int_R \left(\sum_i (f(x_i) - y_i)\,\delta(x - x_i) - \lambda \mathrm{D}^2 f(x) \right) e(x)\mathrm{d}x + \lambda e(x)\mathrm{D}f(x)$$

$$= 0$$

(6.39)

The second term will go to zero as $e(x)$ goes to zero and is not important. However, the term inside the integral might not be zero as $e(x)$ goes to zero. To ensure this, the term in parentheses must be zero, so we are left with

$$0 = \sum_i (f(x_i) - y_i)\,\delta(x - x_i) - \lambda \mathrm{D}^2 f(x)$$ (6.40)

which is a differential equation whose solution is the desired $f(x)$.

One way to solve this differential equation is to make use of a Green's function [xx]. Green's functions are functions that satisfy the following identity:

$$\mathrm{L}G(x, x_i) = \delta(x - x_i)$$ (6.41)

where L is a linear differential operator and G is the Green's function. For our equation, we can use the second derivative as the linear operator to get

$$0 = \sum_i (f(x_i) - y_i)\,\mathrm{D}^2 G(x, x_i) - \lambda \mathrm{D}^2 f(x)$$

$$\mathrm{D}^2 \sum_i (f(x_i) - y_i)\,G(x, x_i) = \mathrm{D}^2 \lambda f(x)$$

(6.42)

Hence,

$$\frac{1}{\lambda} \sum_i (f(x_i) - y_i) \, G(x, x_i) = f(x) \qquad (6.43)$$

Equation 6.43 is a direct solution to 6.40 and states that the function $f(x)$ is a linear combination of Green's functions. Since the weights of the combination depend themselves on $f(x)$, we finally need a small trick to find a solution.

Let us call the weights of the combination $w_i = \frac{1}{\lambda} \left(f(x_i) - y_i \right)$. We can write equation 6.43 as

$$\sum_i w_i G(x, x_i) = f(x) \qquad (6.44)$$

and we can use data points (x_i, y_i) to find the weights. We rewrite this in matrix/vector form as (we cannot use 6.44 directly because our functional is not only the square error, $f(x_j) \neq y_j$):

$$\mathbf{w} = \frac{1}{\lambda}(\mathbf{f} - \mathbf{y})$$
$$\mathbf{Gw} = \mathbf{f} \qquad (6.45)$$

where

$$\mathbf{w} = \begin{bmatrix} w_1 & w_2 & \cdots & w_n \end{bmatrix}^T \qquad (6.46)$$

$$\mathbf{G} = \begin{bmatrix} G(x_1, x_1) & G(x_1, x_2) & \cdots & G(x_1, x_n) \\ G(x_2, x_1) & G(x_2, x_2) & & G(x_2, x_n) \\ & & \ddots & \\ G(x_n, x_1) & G(x_n, x_2) & \cdots & G(x_n, x_n) \end{bmatrix} \qquad (6.47)$$

$$\mathbf{y} = \begin{bmatrix} y_1 & y_2 & \cdots & y_n \end{bmatrix}^T \qquad (6.48)$$
$$\qquad (6.49)$$

whose solution can be obtained by

$$\mathbf{w} = \frac{1}{\lambda}(\mathbf{Gw} - \mathbf{y})$$
$$(\mathbf{G} - \lambda\mathbf{I}) \, \mathbf{w} = \mathbf{y} \qquad (6.50)$$

This finally gives

$$\mathbf{w} = (\mathbf{G} - \lambda\mathbf{I})\,\mathbf{y} \qquad (6.51)$$

6.4 Kernel Mapping (The "Kernel Trick")

Mercer's theorem says that any positive semi-definite function $K(\mathbf{x}_1, \mathbf{x}_2)$, satisfying the relation

$$\sum_{i,j} a_i a_j K(\mathbf{x}_i, \mathbf{x}_j) \geqslant 0 \qquad (6.52)$$

for all pair of vectors $\mathbf{x}_i, \mathbf{x}_j$ and any pair of numbers a_i, a_j, is an inner product in some high-dimensional space, which is called a feature space. This new space is mapped from the space where vectors $\mathbf{x}$ lie to the new high-dimensional space through a non-linear mapping $\Theta : \Re^n \to H$. In another words, there is a non-linear mapping:

$$\mathbf{y} = \phi(\mathbf{x}) \qquad (6.53)$$

such that the inner product in the new mapped space can be computed using some kernel as

$$\langle \mathbf{y}_1, \mathbf{y}_2 \rangle = \langle \phi(\mathbf{x}_1), \phi(\mathbf{x}_2) \rangle = K(\mathbf{x}_1, \mathbf{x}_2) \qquad (6.54)$$

This is also referred to as the "kernel trick". By computing the kernel, which is a simple function to compute, you are indeed using a, possibly very complicated, transformation to high-dimensional space without having to know the transformation.

Examples of kernels are:

- Linear kernels: $K(\mathbf{x}_1, \mathbf{x}_2) = \langle \mathbf{x}_1, \mathbf{x}_2 \rangle$

- Polynomial kernels: $K(\mathbf{x}_1, \mathbf{x}_2) = (\langle \mathbf{x}_1, \mathbf{x}_2 \rangle - 1)^p$

- Gaussian kernels: $K(\mathbf{x}_1, \mathbf{x}_2) = \frac{1}{\sqrt{2\pi\sigma^2}} e^{-\|\mathbf{x}_1 - \mathbf{x}_2\|^2}$

Depending on the kernel, the dimension of the feature space can be arbitrarily large, even infinite. From the analysis, one can show that the input space is mapped to a Hilbert space, and, in some cases, to a reproducing kernel, Hilbert space.

6.5 Classification Problem Using Kernels

Consider the two sets $X = \{\mathbf{x}_1, \mathbf{x}_2, \ldots, \mathbf{x}_N\}$ and $Y = \{\mathbf{y}_1, \mathbf{y}_2, \ldots, \mathbf{y}_N\}$. One wishes to build a classifier that is able to classify a new point $\mathbf{p}$ as belonging to the set X or Y. The simplest way to do so is through the minimal distance classifier. We simply measure the square distance from the point $\mathbf{p}$ to the centre of each class and decide that the point belongs to the class with the smaller square distance. Formally we compute the square distances using inner products as

$$
\begin{aligned}
d_{p,X} &= \langle \mathbf{p} - \mathbf{c}_X, \mathbf{p} - \mathbf{c}_X \rangle \\
d_{p,Y} &= \langle \mathbf{p} - \mathbf{c}_Y, \mathbf{p} - \mathbf{c}_Y \rangle
\end{aligned}
\tag{6.55}
$$

where the centres can be computed as simple averages as

$$
\begin{aligned}
\mathbf{c}_X &= \frac{1}{N} \sum_{i=1}^{N} \mathbf{x}_i \\
\mathbf{c}_Y &= \frac{1}{N} \sum_{i=1}^{N} \mathbf{y}_i
\end{aligned}
\tag{6.56}
$$

Now we can write the distances as (similarly for Y)

$$
d_{p,X} = \left\langle \mathbf{p} - \frac{1}{N} \sum_{i=1}^{N} \mathbf{x}_i, \mathbf{p} - \frac{1}{N} \sum_{i=1}^{N} \mathbf{x}_i \right\rangle
\tag{6.57}
$$

By the linearity property of inner products, we get

$$
d_{p,X} = \frac{1}{N^2} \sum_{i=1}^{N} \sum_{j=1}^{N} \langle \mathbf{p} - \mathbf{x}_i, \mathbf{p} - \mathbf{x}_j \rangle
\tag{6.58}
$$

Distributing the inner products, we have

$$d_{p,X} = \frac{1}{N^2} \sum_{i=1}^{N} \sum_{j=1}^{N} \langle \mathbf{p}, \mathbf{p} \rangle - \langle \mathbf{p}, \mathbf{x}_j \rangle - \langle \mathbf{x}_i, \mathbf{p} \rangle + \langle \mathbf{x}_i, \mathbf{x}_j \rangle$$

$$d_{p,X} = \langle \mathbf{p}, \mathbf{p} \rangle - \frac{2}{N} \sum_{i=1}^{N} \langle \mathbf{p}, \mathbf{x}_i \rangle + b_x$$

(6.59)

where the constant term $b_x = \frac{1}{N^2} \sum_{i=1}^{N} \sum_{j=1}^{N} \langle \mathbf{x}_i, \mathbf{x}_j \rangle$.

Now we build a hypothesis $H_X : d_{p,X} < d_{p,Y}$ for classifying $\mathbf{p}$ to the class X and we get

$$\sum_{i=1}^{N} \langle \mathbf{p}, \mathbf{x}_i \rangle - b_x > \sum_{i=1}^{N} \langle \mathbf{p}, \mathbf{y}_i \rangle - b_y \qquad (6.60)$$

We can now use a kernel to compute the inner product, leading to

$$\sum_{i=1}^{N} K\left(\mathbf{p}, \mathbf{x}_i\right) - b_x > \sum_{i=1}^{N} K\left(\mathbf{p}, \mathbf{y}_i\right) - b_y \qquad (6.61)$$

If the kernel is normalised to integrate to one, the constants b_x and b_y are both one and this result becomes exactly the test using Parzen's windows. Each side corresponds to the probability of $\mathbf{p}$ belonging to the corresponding class. Notice the change in signal due to the negative signal in the distance equation leads to the hypothesis that decides for a larger probability.

6.5.1   Example

Let X and Y be two sets, as follows:

$$X = \{\mathbf{x}_1, \ldots, \mathbf{x}_{10} \in \Re^2 : \|\mathbf{x}_i\| < 1\}$$
$$Y = \{\mathbf{y}_1, \ldots, \mathbf{y}_{10} \in \Re^2 : \mathbf{y}_i \notin X\}$$

(6.62)

To test whether a point $\mathbf{p}$ belongs to the set X, we test the condition either in the input space or in the feature space.

To solve this we need a kernel. Let us use this simple polynomial kernel

$$K(\mathbf{x}_1, \mathbf{x}_2) = \langle \mathbf{x}_1, \mathbf{x}_2 \rangle^2 \tag{6.63}$$

This kernel leads to the following mapping:

$$\begin{bmatrix} x_1 \\ x_2 \end{bmatrix} \rightarrow \begin{bmatrix} u_1 \\ u_2 \\ u_3 \end{bmatrix} = \begin{bmatrix} x_1^2 \\ \sqrt{2}x_1 x_2 \\ x_2^2 \end{bmatrix} \tag{6.64}$$

We can see that the inner product in the new space is

$$\langle \mathbf{u}, \mathbf{v} \rangle = \begin{bmatrix} u_1^2 & \sqrt{2}u_1 u_2 & u_2^2 \end{bmatrix} \begin{bmatrix} v_1^2 & \sqrt{2}v_1 v_2 & v_2^2 \end{bmatrix}^t =$$
$$u_1^2 v_1^2 + 2u_1 v_1 u_2 v_2 + u_2^2 v_2^2 = \langle [\begin{smallmatrix} u_1 & u_2 \end{smallmatrix}], [\begin{smallmatrix} v_1 & v_2 \end{smallmatrix}] \rangle^2 \tag{6.65}$$

This is the kernel inner product we choose. Now let us use this mapping to map the points in each class. First, let us analyse the property that defines the X class:

$$\|\mathbf{x}\| < 1 \tag{6.66}$$

We can rewrite this condition as

$$\langle \mathbf{x}, \mathbf{x} \rangle < 1 \Rightarrow x_1^2 + x_2^2 < 1 \tag{6.67}$$

This is a non-linear boundary for the class X. Now in the feature space

$$\langle \mathbf{u}, \mathbf{u} \rangle < 1 \Rightarrow u_1^2 + u_2^2 + u_3^2 < 1 \tag{6.68}$$

from the mapping we have that

$$u_2^2 = 2u_1 u_3 \tag{6.69}$$

so we have

$$\langle \mathbf{u}, \mathbf{u} \rangle < 1 \Rightarrow u_1^2 + 2u_1 u_3 + u_3^2 < 1$$
$$u_1 + u_3 < 1 \tag{6.70}$$

which is a linear decision boundary. Therefore, the original non-linear boundary in the original space is now a linear one in the feature space.

6.6   Regression Problem Using Kernels

Consider the problem of finding a relationship $y = f(\mathbf{x})$ from certain observations $X = \{\mathbf{x}_0, \mathbf{x}_1, \ldots, \mathbf{x}_N\}$ and $Y = \{y_0, y_1, \ldots, y_N\}$. The simplest possible function consists of a linear relationship of the form

$$y_i = w_1 x_1^i + w_2 x_2^i + \ldots + w_n x_n^i \tag{6.71}$$

where the numbers x_i^j are the j-th components of the n-dimensional vector $\mathbf{x}_i$ and $\mathbf{w} = \begin{bmatrix} w_1 & w_2 & \cdots & w_n \end{bmatrix}^t$ are the parameters of the function. This function can also be written as (using an inner product)

$$y_i = \langle \mathbf{w}, \mathbf{x}_i \rangle \tag{6.72}$$

For each i we have an equation and from the N measurements we can assemble a, usually overdetermined, linear system that can be solved using least squares (linear regression problem):

$$y_1 = \langle \mathbf{w}, \mathbf{x}_1 \rangle$$
$$y_2 = \langle \mathbf{w}, \mathbf{x}_2 \rangle$$
$$\ldots \tag{6.73}$$
$$y_N = \langle \mathbf{w}, \mathbf{x}_N \rangle$$

Now, our goal is to write the least squares problem wholly in terms of inner products of the measurements, so that we can rewrite it as kernels and solve it in feature space. To do this, let us assume that our solution lies in some sub-space of dimension m. Thus, we can choose a linear independent basis

$C = \{c_1, c_2, ..., c_m\}$ to represent this sub-space and write the solution in that sub-space as

$$\mathbf{w} = \sum_{j=1}^{m} a_j \mathbf{c}_j \tag{6.74}$$

Now let us move the problem of finding the solution $\mathbf{w}$ to the regression problem of finding the coordinates $a_j, j = 1, ..., m$. To do so, let us plug the expression of $\mathbf{w}$ into the system of equations. We get

$$y_1 = \left\langle \sum_{j=1}^{m} a_j \mathbf{c}_j, \mathbf{x}_1 \right\rangle$$

$$y_2 = \left\langle \sum_{j=1}^{m} a_j \mathbf{c}_j, \mathbf{x}_2 \right\rangle \tag{6.75}$$

$$\cdots$$

$$y_N = \left\langle \sum_{j=1}^{m} a_j \mathbf{c}_j, \mathbf{x}_N \right\rangle$$

Using the properties of the inner product we get

$$y_1 = \sum_{j=1}^{m} a_j \langle \mathbf{c}_j, \mathbf{x}_1 \rangle$$

$$y_2 = \sum_{j=1}^{m} a_j \langle \mathbf{c}_j, \mathbf{x}_2 \rangle \tag{6.76}$$

$$\cdots$$

$$y_N = \sum_{j=1}^{m} a_j \langle \mathbf{c}_j, \mathbf{x}_N \rangle$$

which in matrix form gives us

$$
\begin{bmatrix}
\langle \mathbf{c}_1, \mathbf{x}_1 \rangle & \langle \mathbf{c}_2, \mathbf{x}_1 \rangle & \cdots & \langle \mathbf{c}_m, \mathbf{x}_1 \rangle \\
\langle \mathbf{c}_1, \mathbf{x}_2 \rangle & \langle \mathbf{c}_2, \mathbf{x}_2 \rangle & & \langle \mathbf{c}_m, \mathbf{x}_2 \rangle \\
\vdots & & \ddots & \\
\langle \mathbf{c}_1, \mathbf{x}_N \rangle & \langle \mathbf{c}_2, \mathbf{x}_N \rangle & & \langle \mathbf{c}_m, \mathbf{x}_N \rangle
\end{bmatrix}
\begin{bmatrix}
a_1 \\ a_2 \\ \vdots \\ a_m
\end{bmatrix}
=
\begin{bmatrix}
y_1 \\ y_2 \\ \vdots \\ a_N
\end{bmatrix}
\tag{6.77}
$$

This is our system written wholly using inner products. Now we use the kernels to compute the inner products to get

$$
\begin{bmatrix}
K\left(\mathbf{c}_1, \mathbf{x}_1\right) & K\left(\mathbf{c}_2, \mathbf{x}_1\right) & \cdots & K\left(\mathbf{c}_m, \mathbf{x}_1\right) \\
K\left(\mathbf{c}_1, \mathbf{x}_2\right) & K\left(\mathbf{c}_2, \mathbf{x}_2\right) & & K\left(\mathbf{c}_m, \mathbf{x}_2\right) \\
\vdots & & \ddots & \\
K\left(\mathbf{c}_1, \mathbf{x}_N\right) & K\left(\mathbf{c}_2, \mathbf{x}_N\right) & & K\left(\mathbf{c}_m, \mathbf{x}_N\right)
\end{bmatrix}
\begin{bmatrix}
a_1 \\ a_2 \\ \vdots \\ a_m
\end{bmatrix}
=
\begin{bmatrix}
y_1 \\ y_2 \\ \vdots \\ a_N
\end{bmatrix}
\tag{6.78}
$$

This is the classical RBF regression solution when we use radial basis functions as kernels. It is also the echo-state solution if we choose kernels at random with soft threshold functions.

6.6.1   Example

Consider the set of measurements shown in figure 6.3(a).

Clearly this is not a linear mapping, so linear regression cannot model the measurements. Let us then use a kernel regression to solve the problem. We will use a Gaussian kernel of the form

$$
K(x_1, x_2) = e^{-3(x_1 - x_2)^2}
\tag{6.79}
$$

Now we solve the regression using kernels. First let us choose a set as basis. Let the basis be

$$
C = \{-1.5, -1, -0.5, 0, 0.5, 1, 1.5\}
\tag{6.80}
$$

Now we assemble the linear system:

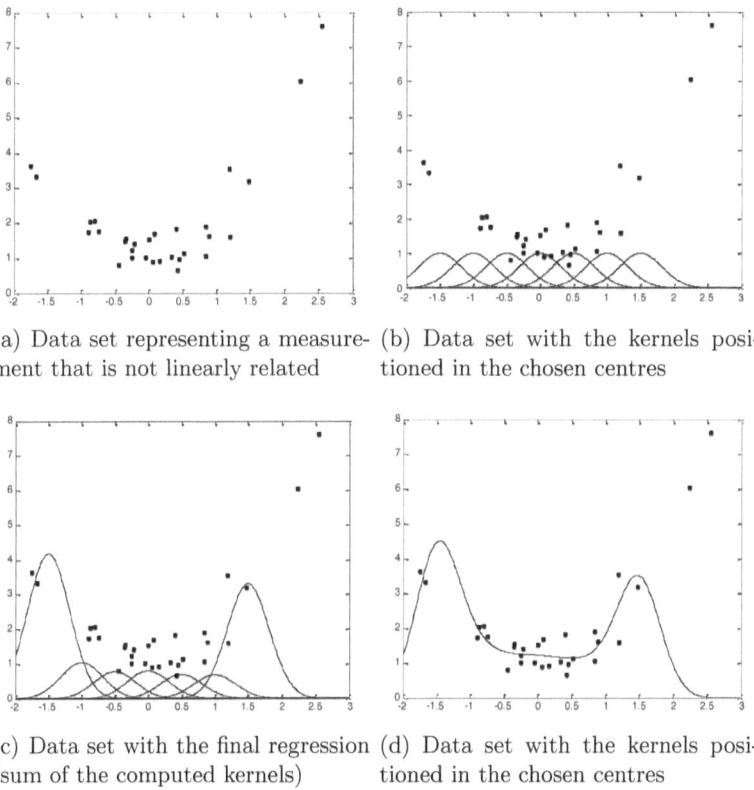

(a) Data set representing a measurement that is not linearly related

(b) Data set with the kernels positioned in the chosen centres

(c) Data set with the final regression (sum of the computed kernels)

(d) Data set with the kernels positioned in the chosen centres

Figure 6.3: Example of regression using kernels

$$
\begin{bmatrix}
K(c_1, x_1) & K(c_1, x_1) & \cdots & K(c_7, x_1) \\
K(c_1, x_2) & K(c_2, x_2) & & K(c_7, x_2) \\
\vdots & & \ddots & \\
K(c_1, x_{30}) & K(c_2, x_{30}) & & K(c_7, x_{30})
\end{bmatrix}
\begin{bmatrix}
a_1 \\
a_2 \\
\vdots \\
a_7
\end{bmatrix}
=
\begin{bmatrix}
y_1 \\
y_2 \\
\vdots \\
y_{30}
\end{bmatrix}
$$

$$(6.81)$$

Using the chosen kernels and basis we have 30 equations of the form

$$a_1 e^{-(x_i - c_1)^2} + a_2 e^{-(x_i - c_2)^2} + \ldots + a_7 e^{-(x_i - c_6)^2} = y_i \quad (6.82)$$

This is a linear combination of Gaussians (shown in figure 6.3(b)). Solving the system and finding the s's, each of the Gaussians is weighted accordingly, resulting in something like figure 6.3(c). Finally if we sum the Gaussians we get the result shown in figure 6.3(d).

Chapter 7

Miscellaneous Topics

$\mathbf{M}$ost of the topics considered up to now were specifically tied to an area or discipline. In this chapter we choose to present a miscellaneous collection of topics that, although they could be tied to one of the previous chapters, are not exactly linked to any of the previous chapters.

7.1 Euler's Formulas (Some of Them)

7.1.1 Euler's Identity

Let us expand the functions $\cos(\varphi t)$ and $\sin(\varphi t)$ using their Taylor series (around 0)

$$\sin(\varpi) = \varpi - \frac{\varpi^3}{3!} + \frac{\varpi^5}{5!} - \frac{\varpi^7}{7!} + \frac{\varpi^9}{9!} \cdots$$
$$\cos(\varpi) = 1 - \frac{\varpi^2}{2!} + \frac{\varpi^4}{4!} - \frac{\varpi^6}{6!} + \frac{\varpi^8}{8!} - \frac{\varpi^{10}}{10!} + \cdots$$

(7.1)

Now we observe that if we add the two series we have all power terms, just like the Taylor series for the exponential. The problem is the changes in sign. Instead of alternating or keeping the sign unchanged, the series resulting in the addition of the two series shows a pattern like $+, +, -, -, +, +, -, -, \ldots$. So Euler must have realised that this pattern of signals is obtained when we have powers of the imaginary unity j as $1, j, -1, -j, 1, j, -1, -j, \ldots$. Hence, we could write the series for a complex exponential as

$$e^{j\varpi} = 1 + j\varpi - \frac{\varpi^2}{2!} - j\frac{\varpi^3}{3!} + \frac{\varpi^4}{4!} + j\frac{\varpi^5}{5!} - \frac{\varpi^6}{6!} - j\frac{\varpi^7}{7!} + \ldots \quad (7.2)$$

which by inspection leads to

$$e^{j\varpi} = \cos(\varpi) + j\sin(\varpi) \quad (7.3)$$

7.1.2   Euler's Totient Function

Euler's totient function $\varphi(n)$ is, in one of its definitions, the number of numbers less than n that are relatively prime to n (that is, they have no common divisor with n). To compute the value of $\varphi(n)$ one must proceed with the following analysis.

The idea is to compute the function $\mathrm{div}(n)$ as the number of numbers that have common divisors with n and use that to compute $\varphi(n)$. As $\varphi(n)$ is the number of numbers that have no common factor with n, all numbers that are smaller than n and are not accounted for by $\mathrm{div}(n)$, must be accounted for by $\varphi(n)$, so

$$\varphi(n) = n - \mathrm{div}(n) \quad (7.4)$$

To compute $\mathrm{div}(n)$, let f_i, $i = 1, 2, 3, \ldots, N$, be the N non-repeating factors of n. For each f_i, we can count all multiples of itself less than n. For example, for $f_i = 3$, we count 3, 6, 9, 12, $\ldots$ (which gives us $n/f_i = n/3$ terms) and if $f_i = 2$ we count 2, 4, 6, 8, 10, 12, $\ldots$ (or $n/2$ terms) and so on. We then notice that we account for 6, 12, $\ldots$ twice (the multiples of the product $f_1 f_2 = 2 \cdot 3 = 6$). This happens with all pair of factors, so we subtract the number of factors of the product of pairs smaller than n. Doing so we now subtract some common pairs twice like 5 times 2×3 and 3 times 5×2 so we sum the multiples of the triples of factors. Then again, we sum extra quadruples, etc. So we arrive at a formula that counts the number of divisors for each factor subtracted from the number of divisors of each pair, etc. This can be written as

$$
\begin{aligned}
\text{div}(n) = {} & \frac{n}{f_1} + \frac{n}{f_2} + \cdots + \frac{n}{f_N} \\
& - \frac{n}{f_1 f_2} - \frac{n}{f_2 f_3} - \cdots - \frac{n}{f_{N-1} f_N} \\
& + \frac{n}{f_1 f_2 f_3} + \frac{n}{f_1 f_2 f_4} + \cdots + \frac{n}{f_{N-2} f_{N-1} f_N} \\
& - \cdots \\
& + \frac{n}{f_2 \cdots f_N} + \frac{n}{f_1 f_3 \cdots f_N} + \cdots \frac{n}{f_1 \cdots f_{N-1}} + \\
& \pm \frac{n}{f_1 \cdots f_N}
\end{aligned}
\tag{7.5}
$$

Summing the fractions we have

$$
\text{div}(n) = \frac{n}{f_1 f_1 \cdots f_N} \left(\begin{array}{c} \sum\limits_{i,j,\cdots,l} f_i f_j \cdots f_l - \cdots \pm \\[2mm] \sum\limits_{i,j,k} f_i f_j f_k \pm \sum\limits_{i,j} f_i f_j \pm \sum\limits_{i} f_i \pm \cdots \pm 1 \end{array} \right)
\tag{7.6}
$$

Notice that the last factors could be plus or minus depending on the number of factors (even or odd). If we have repeated factors, n would be equal to $f_1^{n_1} f_2^{n_2} \cdots f_N^{n_N}$ so we can write

$$
\text{div}(n) = f_1^{n_1-1} f_2^{n_2-1} \cdots f_N^{n_N-1} \times
$$

$$
\left(\sum\limits_{i,j,\cdots,l} f_i f_j \cdots f_l - \cdots \pm \sum\limits_{i,j,k} f_i f_j f_k \pm \sum\limits_{i,j} f_i f_j \pm \sum\limits_{i} f_i \pm \cdots \pm 1 \right)
\tag{7.7}
$$

Now we can use the following identity

$$(f_1-1)(f_2-1)\cdots(f_N-1) = f_1 f_2 \cdots f_N - f_2 \cdots f_N + \cdots \pm f_1 f_3 \cdots f_N \pm \cdots \pm 1 \tag{7.8}$$

So we can write $\text{div}(n)$ as

$$\text{div}(n) = f_1^{n_1-1} f_2^{n_2-1} \cdots f_N^{n_N-1} \left(f_1 \cdots f_N - (f_1-1)(f_2-1)\cdots(f_N-1)\right) \tag{7.9}$$

Now we plug $\text{div}(n)$ into the expression for $\varphi(n)$ and write

$$\begin{aligned}
\varphi(n) &= \\
n &- f_1^{n_1-1} f_2^{n_2-1} \cdots f_N^{n_N-1} \left(f_1 \cdots f_N - (f_1-1)(f_2-1)\cdots(f_N-1)\right) \\
&= f_1^{n_1} f_2^{n_2} \cdots f_N^{n_N} - f_1^{n_1-1} f_2^{n_2-1} \cdots f_N^{n_N-1} \times \\
&\quad (f_1 \cdots f_N - (f_1-1)(f_2-1)\cdots(f_N-1)) \\
&= f_1^{n_1-1} f_2^{n_2-1} \cdots f_N^{n_N-1} \times \\
&\quad (f_1 \cdots f_N - (f_1 \cdots f_N - (f_1-1)(f_2-1)\cdots(f_N-1)))
\end{aligned} \tag{7.10}$$

This leads to

$$\varphi(n) = f_1^{n_1-1} f_2^{n_2-1} \cdots f_N^{n_N-1}(f_1-1)(f_2-1)\cdots(f_N-1) \tag{7.11}$$

So, the celebrated Euler's totient function elegantly computes the number of numbers that have no common divisor with a number n using its factors f_i and the number of repetitions n_i of each factor as

$$\varphi(n) = \prod_i f_i^{n_i-1}(f_i - 1) \tag{7.12}$$

This can be beautifully rewritten as (without explicitly having to write the number of factors n_i)

$$\varphi(n) = n \prod_i \frac{f_i - 1}{f_i} \tag{7.13}$$

7.2   Linear Algebra

7.2.1   Inverse for Matrices with Rank One Terms

One useful formula in basic linear algebra is the inverse of a matrix of the form

$$\mathbf{A} = \mathbf{I} + \mathbf{u}\mathbf{v}^t \tag{7.14}$$

where $\mathbf{I}$ is the identity matrix and $\mathbf{u}$ and $\mathbf{v}$ are vectors of adequate sizes (superscript t means transpose). This formula is interesting because the term $\mathbf{u}\mathbf{v}^t$ has rank equal to one and therefore it does not have an inverse. Nevertheless, when combined with the identity matrix, the sum does have an inverse.

To obtain a formula for the inverse of $\mathbf{A}$ we will proceed as follows. We multiply both sides by $\cong$ and manipulate as follows:

$$\mathbf{A}\mathbf{u} = (\mathbf{I} + \mathbf{u}\mathbf{v}^t)\mathbf{u}$$
$$\mathbf{A}\mathbf{u} = \mathbf{u} + \mathbf{u}\mathbf{v}^t\mathbf{u} \tag{7.15}$$
$$\mathbf{A}\mathbf{u} = (1 + \mathbf{v}^t\mathbf{u})\mathbf{u}$$

Now let us define $k = 1 + \mathbf{v}^t\mathbf{u}$ and multiply both sides by $\mathbf{A}^{-1}$ (assuming it exists)

$$\mathbf{A}\mathbf{u}\mathbf{v}^t = k\mathbf{u}\mathbf{v}^t$$
$$\mathbf{A}^{-1}\mathbf{A}\mathbf{u}\mathbf{v}^t = \mathbf{A}^{-1}k\mathbf{u}\mathbf{v}^t \tag{7.16}$$
$$\mathbf{u}\mathbf{v}^t = \mathbf{A}^{-1}k\mathbf{u}\mathbf{v}^t$$

We now subtract from both sides the term $k\mathbf{I}$ and write the last identity as $\mathbf{A}^{-1}\mathbf{A}$:

$$\mathbf{u}\mathbf{v}^t - k\mathbf{I} = \mathbf{A}^{-1}k\mathbf{u}\mathbf{v}^t - k\mathbf{I}$$
$$\mathbf{u}\mathbf{v}^t - k\mathbf{I} = \mathbf{A}^{-1}k\mathbf{u}\mathbf{v}^t - k\mathbf{A}^{-1}\mathbf{A} \tag{7.17}$$
$$\mathbf{u}\mathbf{v}^t - k\mathbf{I} = k\mathbf{A}^{-1}(\mathbf{u}\mathbf{v}^t - \mathbf{A})$$

and proceed with some algebra to get

$$\frac{1}{k}\mathbf{uv}^t - \mathbf{I} = \mathbf{A}^{-1}(\mathbf{uv}^t - \mathbf{I} - \mathbf{uv}^t)$$

$$\frac{1}{k}\mathbf{uv}^t - \mathbf{I} = \mathbf{A}^{-1}(-\mathbf{I})$$

(7.18)

which finally leads to

$$\mathbf{I} - \frac{1}{1+\mathbf{v}^t\mathbf{u}}\mathbf{uv}^t = \mathbf{A}^{-1} \tag{7.19}$$

7.2.2 Exponential of a Matrix

We wish to find a way to compute the following expression for the exponential of a matrix $\mathbf{A}$

$$e^{\mathbf{A}} \tag{7.20}$$

To do so, we will investigate the solution for the following vector differential equation

$$\dot{\mathbf{x}}(t) = \mathbf{A}\mathbf{x}(t) \tag{7.21}$$

We start by writing the solution in its Taylor series representation as

$$\mathbf{x}(t) = \mathbf{c}_0 + \mathbf{c}_1 t + \frac{\mathbf{c}_2 t^2}{2!} + \frac{\mathbf{c}_3 t^3}{3!} + \cdots \tag{7.22}$$

If we differentiate this with respect to t we get

$$\dot{\mathbf{x}}(t) = \mathbf{c}_1 + \mathbf{c}_2 t + \frac{\mathbf{c}_3 t^2}{2!} + \frac{\mathbf{c}_4 t^3}{3!} + \cdots \tag{7.23}$$

Now using 7.21 we have the equality

$$\mathbf{A}\mathbf{c}_0 + \mathbf{A}\mathbf{c}_1 t + \frac{\mathbf{A}\mathbf{c}_2 t^2}{2!} + \frac{\mathbf{A}\mathbf{c}_3 t^3}{3!} + \cdots =$$

$$\mathbf{c}_1 + \mathbf{c}_2 t + \frac{\mathbf{c}_3 t^2}{2!} + \frac{\mathbf{c}_4 t^3}{3!} + \cdots \tag{7.24}$$

and equating the coefficients with the same power we have

$$\mathbf{c}_1 = \mathbf{A}\mathbf{c}_0$$
$$\mathbf{c}_2 = \mathbf{A}\mathbf{c}_1$$
$$\mathbf{c}_3 = \mathbf{A}\mathbf{c}_2 \tag{7.25}$$
$$\vdots$$

This can be developed to

$$\mathbf{c}_1 = \mathbf{A}\mathbf{c}_0$$
$$\mathbf{c}_2 = \mathbf{A}^2\mathbf{c}_0$$
$$\mathbf{c}_3 = \mathbf{A}^3\mathbf{c}_0 \tag{7.26}$$
$$\vdots$$

Now we can write the series for $\mathbf{x}(t)$ as

$$\mathbf{x}(t) = \mathbf{c}_0 \left(\mathbf{I} + \mathbf{A}\mathbf{c}_1 t + \frac{(\mathbf{A}t)^2}{2!} + \frac{(\mathbf{A}t)^3}{3!} + \cdots \right)$$
$$= \mathbf{c}_0 e^{\mathbf{A}t} \tag{7.27}$$

and making $t = 0$ we have that $\mathbf{c}_0 = \mathbf{x}(0)$. Now we have a relationship between $e^{\mathbf{A}}$ and the solution of 7.21. But it is possible to solve 7.21 analytically using the Laplace transform. We just apply the transformation to both sides and get

$$L\{\dot{\mathbf{x}}(t)\} = L\{\mathbf{A}\mathbf{x}(t)\}$$
$$s\mathbf{x}(s) - \mathbf{x}(0) = \mathbf{A}\mathbf{x}(s) \tag{7.28}$$

which leads to

$$\mathbf{x}(s) = \mathbf{x}(0)(s\mathbf{I} - \mathbf{A})^{-1} \tag{7.29}$$

Now computing the inverse transform we get

$$\mathbf{x}(s) = \mathbf{x}(0)L^{-1}\left\{(s\mathbf{I} - \mathbf{A})^{-1}\right\}$$
$$= \mathbf{x}(0)e^{\mathbf{A}t} \tag{7.30}$$

Hence, the formula for computing the exponential of a matrix is

$$e^{\mathbf{A}} = L^{-1}\left\{(s\mathbf{I} - \mathbf{A})^{-1}\right\}\Big|_{t=1} \tag{7.31}$$

7.3 Paradoxes

7.3.1   4 = 5

Let us start with an identity

$$a = b \tag{7.32}$$

Now, multiply both sides by 4

$$4a = 4b \tag{7.33}$$

We could add b to both sides

$$4a + b = 5b \tag{7.34}$$

Now, subtract $5a$ from both sides. Since $a = b$ we use $5a$ on the right and $5b$ on the left

$$
\begin{aligned}
4a + b - 5b &= 5b - 5a \\
4a - 4b &= 5b - 5a
\end{aligned} \tag{7.35}
$$

Isolating 4 on the left and 5 on the right we get

$$4(a - b) = 5(a - b) \tag{7.36}$$

This simplifies to

$$4 = 5 \tag{7.37}$$

7.3.2 Imaginary Numbers

We can write a trivial equality with the imaginary unity i simply as

$$i = i \tag{7.38}$$

Using the definition for the imaginary number as $i = \sqrt{-1}$ we have

$$\sqrt{-1} = \sqrt{-1} \tag{7.39}$$

We now perform a set of algebraic manipulations to the negative unity inside the square roots

$$\sqrt{\frac{-1}{1}} = \sqrt{\frac{-1}{1}}$$
$$\sqrt{\frac{-1}{1}} = \sqrt{\frac{1}{-1}} \tag{7.40}$$
$$\frac{\sqrt{-1}}{1} = \frac{1}{\sqrt{-1}}$$

Now we use the definition for i

$$\frac{i}{1} = \frac{1}{i} \tag{7.41}$$

which by cross multiplication leads us to

$$i^2 = 1$$
$$-1 = 1 \tag{7.42}$$

7.3.3 Any Length Straight Line

This paradox involves the shortest path linking two points. As is thoroughly given in all textbooks, that path is a straight line. To measure the length of the line, we can use Pythagoras' theorem and calculate the length of the hypotenuse of a triangle formed by linking points A and B with the alternative path that is constant in x and then in y (figure 7.1). For instance, if the horizontal base has length equal to 3 and the vertical side has length 4, the path will have size $\sqrt{3^2 + 4^2} = 5$.

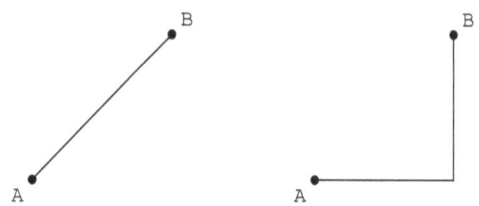

(a) Straight line from A to B

(b) Alternative path from A to B

Figure 7.1: Examples of two paths from point A to point B

Now let us say we measure the length of the path using the alternative horizontal and vertical lines. Now the length is $4 + 3 = 7$. But since that is not the shortest path, we perform a "folding" to approximate it to the real shortest one (as illustrated in figure 7.2(a)). Now the size is the sum of the horizontal and vertical lines: $3/2 + 3/2 + 4/2 + 4/2 = 7$. We can keep folding the lines as illustrated by figures 7.2(b) and 7.2(c) and will keep getting 7 as the length of the alternative path. We will get 7 even in the limit when we fold it so many times that all the points in the alternative path lie on the original shortest path. However, the length will still be 7 (not 5).

7.4   Summations

7.4.1   Arithmetic Progressions

An arithmetic progression is a sequence with N elements of the form

$$\{a_0, a_1 + r, a_1 + 2r, a_1 + 3r, \cdots, a_1 + N\,r\} \qquad (7.43)$$

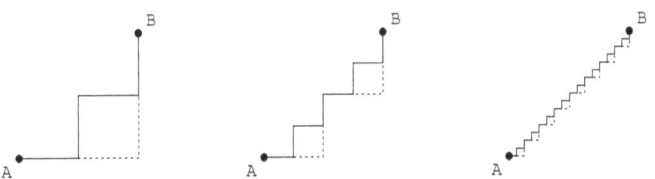

(a) First folding for al- (b) Second folding for (c) Fourth folding for
ternative path   alternative path   alternative path

Figure 7.2: Examples of folding the alternative paths from point
A to point B

where a_0 is the first element and the sequence is formed by
adding r to the previous element up to $a_N = a_1 + N r$.

If we add the first and last elements we get $2 a_1 + N r$. But if
we add the second and the N-th we also have $2 a_1 + N r$. Even if
N is odd, if we add the middle term with itself we get $2 a_1 + N r$.
So, to sum all the terms, we just sum $N/2$ additions of the first
and last elements, therefore the general formula is

$$S = (a_1 + a_N)\frac{N}{2} \tag{7.44}$$

7.4.2 Infinite Geometric Progressions

Consider the geometric sequence

$$\{1, r, r^2, r^3, \cdots\} \tag{7.45}$$

with $|r| < 1$ being the common ratio. We wish to compute the
summation of all infinite elements of this series.

To do so, we write (since the terms will decrease to 0)

$$S = 1 + r + r^2 + r^3 + r^4 + \cdots$$
$$S = 1 + r(1 + r + r^2 + r^3 + \cdots) \tag{7.46}$$

Now we identify the term on the r.h.s. as the summation itself and write

$$S = 1 + rS$$
$$S - rS = 1 \tag{7.47}$$

which finally gives us the formula for the summation of the geometric progression as

$$S = \frac{1}{1 - r} \tag{7.48}$$

7.4.3   Finite Geometric Progressions

We can write a general formula for the summation of the first N elements of a geometric progression. Let

$$S_N = 1 + r + r^2 + r^3 + r^4 + \cdots + r^N \tag{7.49}$$

We can write another sequence as

$$S_{N+1} = r^{N+1} + r^{N+2} + \cdots$$
$$= r^{N+1}S \tag{7.50}$$

The sum of the whole progression is then

$$S = 1 + r + r^2 + r^3 + \cdots$$
$$= S_N + S_{N+1} \tag{7.51}$$

Using the formula for S we have

$$\frac{1}{1 - r} = S_N + \frac{r^{N+1}}{1 - r} \tag{7.52}$$

This leads to the formula for the finite series summation

$$S_N = \frac{1 - r^{N+1}}{1 - r} \tag{7.53}$$

7.4.4 $\sum_{n=0}^{\infty} n\, r^n$

Let us "open" the summation as

$$S = 0 + r + 2r^2 + 3r^3 + 4r^4 + \cdots \tag{7.54}$$

Now let us split off one term of each power of r like

$$\begin{aligned} S &= 0 + r + (r^2 + r^2) + (r^3 + 2r^3) + \\ &\quad (r^4 + 3r^4) + \cdots \\ &= (r + r^2 + r^3 + r^4 + \cdots) + \\ &\quad (r^2 + 2r^3 + 3r^4 + \cdots) \end{aligned} \tag{7.55}$$

Now for the second term we can isolate r from each term and get

$$\begin{aligned} S &= (r + r^2 + r^3 + r^4 + \cdots) + \\ &\quad r(r + 2r^2 + 3r^3 + \cdots) \end{aligned} \tag{7.56}$$

The two terms now are identifiable. The first is a simple geometric progression summation and the second is S itself. Hence we have

$$S = \frac{r}{1-r} + rS \tag{7.57}$$

which solving for S gives us

$$S = \frac{r}{(1-r)^2} \tag{7.58}$$

7.4.5   Inverse Formula Using a Series for a Matrix

Here we would like to investigate the following expression

$$\mathbf{M} = \frac{\mathbf{I}}{s} + \frac{\mathbf{A}}{s^2} + \frac{\mathbf{A}^2}{s^3} + \frac{\mathbf{A}^3}{s^4} + \cdots \tag{7.59}$$

This is the matrix equivalent of the geometric progression, so we will proceed as such. First we rewrite the expression as

$$\mathbf{M} = \frac{\mathbf{I}}{s} + \frac{\mathbf{A}}{s}\left(\frac{\mathbf{I}}{s} + \frac{\mathbf{A}}{s^2} + \frac{\mathbf{A}^2}{s^3} + ...\right) \qquad (7.60)$$

Then we identify the term in parentheses as $\mathbf{M}$ and write

$$s\mathbf{M} = \mathbf{I} + \mathbf{AM}$$
$$s\mathbf{M} - \mathbf{AM} = \mathbf{I} \qquad (7.61)$$

$$\mathbf{M} = \frac{\mathbf{I}}{s} + \frac{\mathbf{A}}{s}\mathbf{M}$$
$$s\mathbf{M} - \mathbf{AM} = \mathbf{I} \qquad (7.62)$$

which leads to

$$(s\mathbf{I} - \mathbf{A})\,\mathbf{M} = \mathbf{I} \qquad (7.63)$$

and finally we have

$$\mathbf{M} = (s\mathbf{I} - \mathbf{A})^{-1} \qquad (7.64)$$

Appendix A

Differential Inclusions

In 1988, Filippov proposed a link between solutions for differential equations with a discontinuous right-hand side (DRHS Dif. Eq.) and differential inclusions. This was immediately applied to dynamical systems with a relay by Utkin and it became a very powerful tool in robust control theory.

A.1 System with Discontinuous Right-hand Side

Following Fillipov, consider the system of order n defined, in general, by

$$\dot{\mathbf{x}} = f(\mathbf{x}) \qquad \text{(A.1)}$$

where $\mathbf{x} \in \Re^n$ is a state vector and $f : \Re^n \to \Re^n$ a discontinuous function of the states. We are interested in the analysis of systems of the form

$$\dot{\mathbf{x}} = \mathbf{A}\mathbf{x} + \mathbf{b}\operatorname{sign}(\mathbf{v}^t\mathbf{x}) \qquad \text{(A.2)}$$

where $\mathbf{A} \in \Re^{n \times n}$ is the transition matrix, $\mathbf{b}, \mathbf{v} \in \Re^n$ are vectors, and the function sign(.) is the sign function defined as

$$\text{sign}(x) = \begin{cases} 1, & x > 0 \\ -1, & x < 0 \\ 0, & x = 0 \end{cases} \tag{A.3}$$

In this manner, in the semi-planes defined by $\mathbf{v}^t\mathbf{x} > 0$ and $\mathbf{v}^t\mathbf{x} < 0$ the system shows linear behaviours of the form

$$\dot{\mathbf{x}} = \mathbf{A}\mathbf{x} + \mathbf{b}$$
$$\dot{\mathbf{x}} = \mathbf{A}\mathbf{x} - \mathbf{b} \tag{A.4}$$

Figure A.1 illustrates an example of such systems. Notice that in the pointed semi-planes, the behaviour is linear. On the other hand, also notice that in the surface given by $\mathbf{v}^t\mathbf{x} = 0$ (denoted as a sliding surface) the behaviour cannot directly be defined since the right-hand side of (A.2) is discontinuous. One way of illustrating this difficulty is to try to apply the value (even with it being discontinuous) of (A.2) for the sliding surface. If we do this, we obtain a system of the form $\dot{\mathbf{x}} = \mathbf{A}\mathbf{x}$. If this were really the behaviour of the system on the sliding surface, it would have an analytical and continuous solution at that point, which contradicts the uniqueness and existence theorems for solutions of differential equations. Intuitively, we notice that the behaviour of the system seems to "switch" from one side to the other when its states fall onto the sliding surface. Also, under certain conditions the states will actually "slide" on this surface. In this section, we are going to show exactly why, when and how the states slide.

Utkin showed that the dynamics for the sliding surface can be obtained using a principle called equivalent control. Basically one uses the fact that, in reaching the sliding surface, the states will "slide" over the surface, making the vector $\dot{\mathbf{x}}$ perpendicular to the vector $\mathbf{v}$. This technique consists in finding a signal u that when applied to the system given by

$$\dot{\mathbf{x}} = \mathbf{A}\mathbf{x} + \mathbf{b}u \tag{A.5}$$

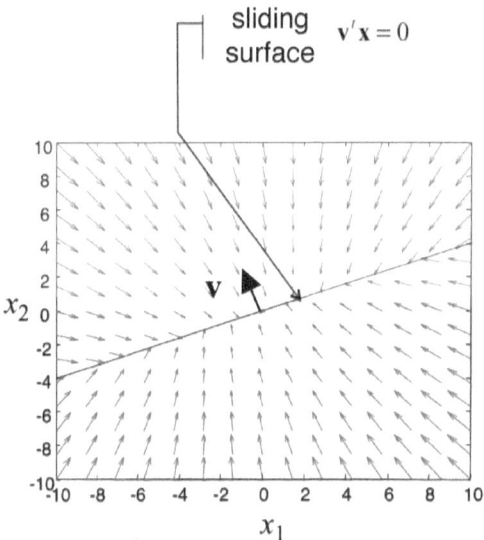

Figure A.1: Example of a system of order two defined by (A.2). The arrows indicate the values of $\dot{\mathbf{x}}$. The filled line that crosses the plot, defining two semi-planes, is the sliding surface. The filled arrow in the middle is the vector $\mathbf{v}$.

gives a trajectory that is always perpendicular to $\mathbf{v}$. This is done by multiplying (A.5) by $\mathbf{v}^t$ and isolating u. Hence, we obtain

$$0 = \mathbf{v}^t \mathbf{A} \mathbf{x} + \mathbf{v}^t \mathbf{b} u$$

$$u = -\frac{\mathbf{v}^t \mathbf{A} \mathbf{x}}{\mathbf{v}^t \mathbf{b}} \qquad (A.6)$$

The zero on the left-hand side is a consequence of the fact that $\mathbf{v}^t$ is perpendicular to $\dot{\mathbf{x}}$. Using this result in (A.5) we get

$$\dot{\mathbf{x}} = \left(\mathbf{I} - \frac{\mathbf{b} \mathbf{v}^t}{\mathbf{v}^t \mathbf{b}} \right) \mathbf{A} \mathbf{x} \qquad (A.7)$$

This is the result that we will obtain in the rest of this section by utilising Filippov theory in a more formal, but illustrative, sense. The simplicity in obtaining such dynamics (using the equivalent control principle) was only possible when we assumed that the trajectory at the sliding surface will correspond to the "sliding" of the states on that surface. This fact (the sliding of the states), although visually obvious in some second-order systems, can only be demonstrated using Filippov theory. Besides, the result itself clarifies neither the limits of the sliding surface nor the conditions for it to exist. In this section we will explore these questions in the light of Filippov theory. We will try to explore the consequences of the so-called principle of the equivalent control, aiming at a more rigorous and exploratory analysis.

A.2   Differential Inclusions

Differential inclusions are equations of the kind

$$\dot{\mathbf{x}} \in F(\mathbf{x}) \tag{A.8}$$

where F is called a multi-function or multivalued function. These kinds of function are defined as

$$F : \Re^n \to C \tag{A.9}$$

where C is the set of all subsets of $\Re^n$. In other words, a multivalued function is a function that gives us a set instead of a unique value. Figure A.2 illustrates a plot of a multivalued function given by $F(x) = [0, x]$.

A differential inclusion is an inequality that relates the differential of the states of a dynamical system through a multivalued function as given in (A.8). A trajectory starting from $\mathbf{x}(0) = \mathbf{x}_0$ is said to satisfy the differential inclusion if for all times $t > 0$, its derivative obeys the inequality that defines the differential inclusion. Formally we have

$$\mathbf{x}(t) : \Re^+ \to \Re^n | \mathbf{x}(0) = \mathbf{x}_0, \quad \forall t \geqslant 0, \quad \dot{\mathbf{x}}(t) \in F(\mathbf{x}(t)) \tag{A.10}$$

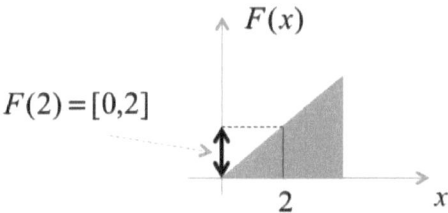

Figure A.2: Example of a multivalued function. The "filled" plot represents the multiple values of the function for each input argument value. One can observe that $F(2)$ has an interval $[0; 2]$, which is a subset of $\Re$.

To illustrate this definition, consider as an example the first-order differential inclusion formed by the multivalued function

$$F(x) = [-0.5x, -x] \tag{A.11}$$

The initial condition is given by $x(0) = 1$. We then have

$$\dot{x} \in [-0.5x, -x]$$
$$x(0) = 1 \tag{A.12}$$

There are infinite trajectories that satisfy the differential inclusion in this example. Figure A.3 illustrates the trajectories that satisfy this differential inclusion.

Notice that there are limiting trajectories where the derivative has maximum and minimum values for the differential inclusion. Any trajectory outside these limits does not satisfy the inclusion for all times and therefore is not a possible trajectory. Importantly, not all trajectories within the limits are possible trajectories. There are some trajectories, like the one indicated in the plot, that, despite being inside the limits, have derivatives that are outside the limits of the differential inclusion. In the example, notice that this trajectory has several points where the tangent is zero (the flat region) and that this value (zero) is outside the limits of the differential inclusion.

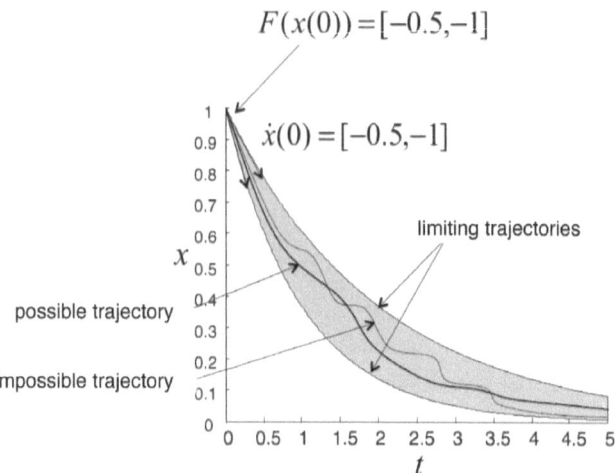

Figure A.3: Illustration of the solutions that satisfy the differential inclusion $\dot{x} \in [-0.5x, -x]$ with initial condition equal to 1. The "filled" part of the plot represents the place where the trajectories must be.

Next, a series of definitions related to the theory of differential inclusions will be presented. They will be used to derive solutions for differential equations with a discontinuous right-hand side having the form of (A.2).

A.2.1   Definitions and Relations Between Differential Inclusions and Solutions to Differential Equations with Discontinuous Right-hand Sides

To use the theory of multivalued functions $F(\mathbf{x}_0) = [v_1, v_2]$ in the solution of a DRHS Dif. Eq., the multivalued function must satisfy three conditions at the discontinuity point $\mathbf{x}_0$:

- Closed in $\mathbf{x}_0$: Formally this condition establishes that, at the point of discontinuity, the interval that is the result of the discontinuity must be closed.

- Convex in $\mathbf{x}_0$: At the point of discontinuity the value of the multivalued function must be a set or a interval that is convex. Formally we have

$$F(\mathbf{x}_0) \subseteq \overline{\mathrm{conv}}(F(\mathbf{x}_0)) \tag{A.13}$$

where $\overline{\mathrm{conv}}(.)$ represents the convex hull of the set.

- Upper semi-continuous in $\mathbf{x}_0$: A multivalued function is considered upper semi-continuous in $\mathbf{x}_0$ if when $\mathbf{x}$ tends to $\mathbf{x}_0$ them $F(\mathbf{x}) \subset F(\mathbf{x}_0)$.

At this point it is possible to establish the relation between differential inclusions and the DRHS Diff. Eq. According to Filippov, the solution of the DRHS Diff. Eq. must obey a differential inclusion at its discontinuity point (for relay systems, the discontinuity is represented by the sign function) and must be replaced, at that point, by a multivalued function given by

$$F(\mathbf{x}) = \bigcap_{\delta > 0} \overline{\mathrm{conv}} \left(f(B(\mathbf{x}, \delta)) \right) \tag{A.14}$$

where $f(.)$ is the original DRHS Diff. Eq. and $B(\mathbf{x}, \delta)$ is a ball of radius δ centred at B. Moreover, $F(\mathbf{x})$ must be closed, convex and upper semi-continuous.

As shown in this section, there are infinite trajectories $\mathbf{x}(t)$ that satisfy the differential inclusion. Not all of them are also solutions for the original DRHS Diff. Eq. Also following Filippov, for a trajectory to be a solution of the original DRHS Diff. Eq., it must satisfy the following condition:

$$\begin{aligned} &\mathbf{x}(t) \in \Re^n, \quad t \in \Re^+ | \mathbf{x}(t) \in \overline{\mathrm{conv}}(F(\mathbf{x}(t))) \\ &\dot{\mathbf{x}}^t(t) s_n(\mathbf{x}(t)) = 0 \text{ if } s(\mathbf{x}(t)) = 0 \end{aligned} \tag{A.15}$$

Here, $s(\mathbf{x}(t)) = 0$ represents the "sliding surface" (the states $\mathbf{x}(t)$ that satisfy $s(\mathbf{x}(t)) = 0$ are the ones that will compose the surface). $s_n(\mathbf{x}(t)) = 0$ are the vectors normal to the sliding surface. In other words, the trajectories that are tangent to the sliding surface and satisfy the differential inclusion during that time, will be solutions to the original DRHS Diff. Eq. in (A.2).

A.2.2   Example

Using the concepts presented in 2.2.1, we can find the multivalued function that is the sign function in the differential inclusion associated with the problem in (A.2). Figure A.4 shows the plot of the sign function as defined in (A.3).

$$f(x) = \begin{cases} -1, & x < 0 \\ 1, & x > 0 \\ 0, & x = 0 \end{cases}$$

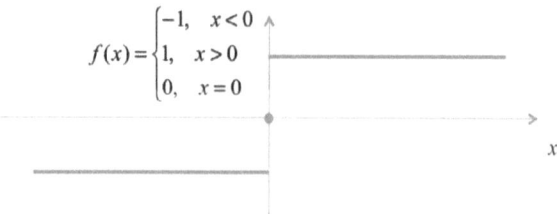

Figure A.4: Plot for the sign function as defined in (A.3).

Equation (A.14) basically establishes, for the sign function, that points outside the discontinuity will result in a set of measure zero with the value of the function (-1 or 1). We can informally state that, observing that the result $S_k = \overline{\text{conv}}\,(f(B(x_0, \delta_k)))$ for $\delta_1 < \delta_2 < \cdots < \delta_i < \delta_{i+1} \cdots$, there is the following tendency

$$
\begin{aligned}
S_1 &= \{-1\} \\
S_2 &= \{-1\} \\
&\;\vdots \\
S_i &= \{-1, 0, 1\} \\
&\;\vdots
\end{aligned}
\qquad\qquad (A.16)
$$

Observe that, in doing the intersection of all convex hulls, the result is the set of measure zero with -1 as the only element. The convex hulls of each set are given by

$$\overline{\text{conv}}(S_1) = \{-1\}$$
$$\overline{\text{conv}}(S_2) = \{-1\}$$
$$\vdots \tag{A.17}$$
$$\overline{\text{conv}}(S_i) = [-1, 1]$$
$$\vdots$$

and their intersection is $\bigcap\limits_{\delta>0} \overline{\text{conv}}\left(f(B(x_0, \delta))\right) = \{-1\}$.

Figure A.5 illustrates the process. For visualisation purposes, only some discrete radius δ's of the balls are shown. Nevertheless, the intersection is for a continuum of radii.

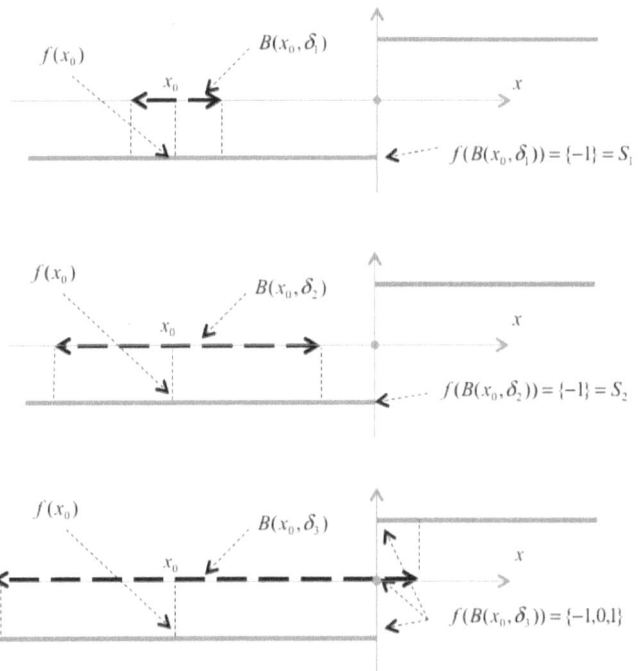

Figure A.5: Illustration of representing the sign function with its equivalent multivalued function using equation (A.14) outside the discontinuity.

With the same procedure, but now observing what happens at the discontinuity, we obtain the following sets

$$S_1 = \{1, 0, -1\}$$
$$S_2 = \{1, 0, -1\}$$
$$\vdots$$
$$S_i = \{1, 0, -1\}$$
$$\vdots$$

(A.18)

with convex hulls given by

$$\overline{\text{conv}}(S_1) = [-1, 1]$$
$$\overline{\text{conv}}(S_2) = [-1, 1]$$
$$\vdots$$
$$\overline{\text{conv}}(S_i) = [-1, 1]$$
$$\vdots$$

(A.19)

This results in the set $\bigcap_{\delta>0} \overline{\text{conv}}\left(f(B(x_0, \delta))\right) = [-1, 1]$. Figure A.6 illustrates the process.

Therefore, the result of the application of the definition (A.14) to the sign function is the multivalued function given by

$$\text{SIGN}(y) = \begin{cases} 1, & y > 0 \\ -1, & y < 0 \\ [-1, 1], & y = 0 \end{cases}$$

(A.20)

Figure A.7 shows the plot of the function represented in (A.20).

A.3   Filippov Local Solution to the Relay System

In this section we will consider the system

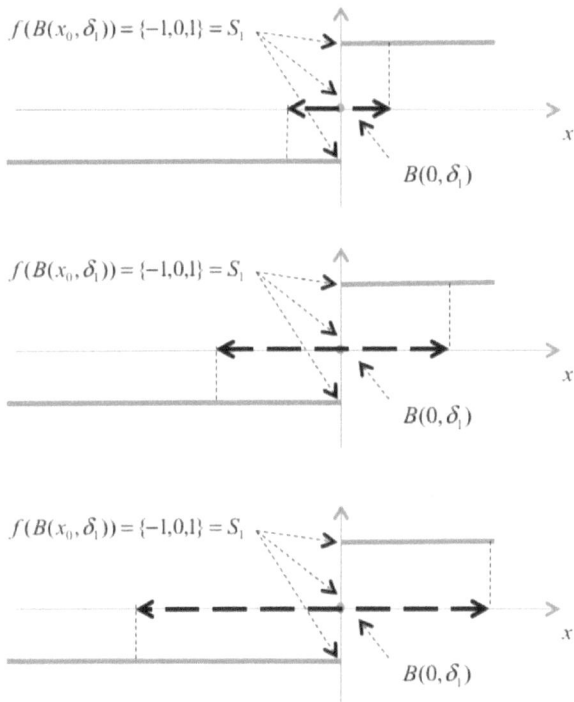

Figure A.6: Illustration of representing the sign function with its equivalent multivalued function at the discontinuity.

$$\dot{\mathbf{x}} = \mathbf{A}\mathbf{x} + \mathbf{b}\,\mathrm{SIGN}\left(\mathbf{v}^t\mathbf{x}\right) \qquad (A.21)$$

As we saw in the previous sections, the differential inclusion that represents this system is given by

$$\dot{\mathbf{x}} \in \mathbf{A}\mathbf{x} + \mathbf{b}\,\mathrm{SIGN}\left(\mathbf{v}^t\mathbf{x}\right) \qquad (A.22)$$

where the multivalued function SIGN(.) is given by (A.20).

As stated by Filippov [filippov], the derivative of the solution of the differential inclusion, at the points of discontinuity of (A.2), must belong to the convex hull formed by the values

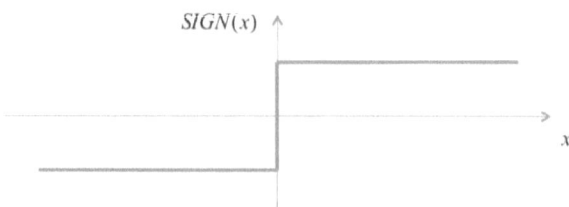

Figure A.7: Plot of the multivalued function defined in (A.20).

of the differential inclusion on the sliding surface. For equation (A.2), we have that $\dot{\mathbf{x}}$ must satisfy

$$\dot{\mathbf{x}} \in \overline{\mathrm{conv}}(C) | C = \{\dot{\mathbf{x}}_1, \dot{\mathbf{x}}_2\} \tag{A.23}$$

where $\dot{\mathbf{x}}_1$ and $\dot{\mathbf{x}}_2$ are the limits of the derivative of the interval of the solution of the differential inclusion given by

$$\begin{aligned} \dot{\mathbf{x}}_1 &= \mathbf{A}\mathbf{x} + \mathbf{b} \\ \dot{\mathbf{x}}_2 &= \mathbf{A}\mathbf{x} - \mathbf{b} \end{aligned} \tag{A.24}$$

There are infinite trajectories that meet this condition. Among them, the trajectory that is a solution of (A.2) must satisfy the conditions (A.15). Next, the definitions necessary to establish such conditions will be presented.

A.4 Local Solution and Equivalent Control

In this section, we will examine the behaviour of the trajectories that are solutions of (A.22) and satisfy condition (A.15). These trajectories are considered to be solutions of (A.2). The first rule is that the trajectory (that will be a solution of (A.2)) must belong to the convex hull of $F(\mathbf{x})$. These trajectories are given by

$$\dot{\mathbf{x}} = \lambda \dot{\mathbf{x}}_1 + (1 - \lambda)\dot{\mathbf{x}}_2$$
$$0 \leqslant \lambda \leqslant 1 \tag{A.25}$$

where $\mathbf{x}_1$ and $\mathbf{x}_2$ are limiting trajectories of $F(\mathbf{x})$ as presented in (A.24). Substituting (A.24) into (A.25) (and changing the variables back to $\mathbf{x}$), we obtain:

$$\dot{\mathbf{x}} = \lambda\,(\mathbf{Ax} + \mathbf{b}) + (1 - \lambda)\,(\mathbf{Ax} - \mathbf{b}) \tag{A.26}$$

This leads to

$$\dot{\mathbf{x}} = \mathbf{Ax} - (1 - 2\lambda)\mathbf{b} \tag{A.27}$$

In this way, by varying λ from 0 to 1, we obtain all the trajectories that comply with the first rule. The second rule is that for the trajectory to be a solution of (A.2) then, from all the trajectories that we obtain by varying λ in (A.27), we must pick the one that is a tangent to the sliding surface. That condition is expressed by

$$\lambda_f = \{\lambda | \mathbf{v}^t \dot{\mathbf{x}} = 0\} \tag{A.28}$$

Applying (A.27) we obtain

$$\mathbf{v}^t\,(\mathbf{Ax} - (1 - 2\lambda_f)\mathbf{b}) = 0 \tag{A.29}$$

This leads to

$$\lambda_f = \frac{1}{2} - \frac{\mathbf{v}^t \mathbf{Ax}}{2\mathbf{v}^t \mathbf{b}} \tag{A.30}$$

This is the value of λ that makes (A.27) a solution of (A.2). Substituting (A.30) into (A.27) we obtain

$$\dot{\mathbf{x}} = \left(\mathbf{I} - \frac{\mathbf{b}\mathbf{v}^t}{\mathbf{v}^t \mathbf{b}}\right) \mathbf{Ax} \tag{A.31}$$

Equation (A.31) represents the dynamics of (A.2) when $\mathbf{x}$ is on the sliding surface. In other words, it represents the dynamics that will hold for the system when its states reach the sliding surface. This solution is called the equivalent control solution. Figure (A.8) illustrates how the dynamics is obtained using (A.15).

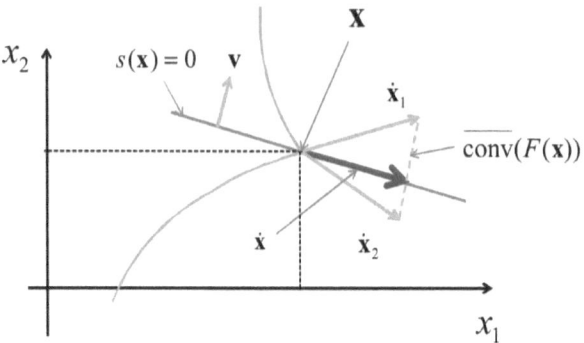

Figure A.8: Representation of the behaviour of a trajectory outside and inside the sliding surface.

The figure represents a system in which the dynamics is given by (A.2). $\dot{\mathbf{x}}$ is the Filippov's solution. Notice that the conditions (A.15) are met, namely, $\dot{\mathbf{x}}$ belongs to the convex hull of $F(\mathbf{x})$ and is tangent to the sliding surface.

A.5   Analysis of Filippov's Solution

Now we will analyse some important aspects of the dynamics of (A.2). In general, the system behaves like one of the systems described in (A.24) when the states are outside the sliding surface. When the states reach the surface given by $\mathbf{v}^t\mathbf{x} = 0$, the system has the dynamics given by (A.31) and some very interesting situations can occur. The first situation is a hard change in velocity but without any sliding. For sliding to occur, the states must be inside certain limits (shown next). Some questions are important when dealing with sliding:

- How does the sliding surface dynamics depend on the relay amplitude?

- What are the sliding limits? Are they stable?

- What are the conditions for the sliding surface to be unbounded?

- Is it possible for the state to go out of the sliding surface? Once it has been reached?

- How does this exit occur?

A.5.1 Relay Dependency

The sliding surface dynamics is given by (A.31). Looking at the dynamics, we can observe that if the relay amplitude $\mathbf{b}$ changes (say to $a\,\mathbf{b}$), representing a change in the relay signal, the term $\frac{\mathbf{b}\mathbf{v}^t}{\mathbf{v}^t\mathbf{b}}$ does not change. Despite being a mathematically trivial result, this has practical consequences, which are very interesting. Basically, the dynamics for the sliding surface does not depend on the relay amplitude and a large value of the relay signal does not imply a faster movement into the sliding surface.

A.5.2 Stability

The stability of the dynamics for the sliding surface is determined by the eigenvalues of $\left(\mathbf{I} - \frac{\mathbf{b}\mathbf{v}^t}{\mathbf{v}^t\mathbf{b}}\right)\mathbf{A}$. The system will be stable (for the sliding surface) if the real part of the eigenvalues is negative. In other words, for the sliding surface, the system behaves like a simple linear system.

A.5.3 Sliding Limits

As we mentioned before, the states may slide on the sliding surface given by $\mathbf{v}^t\mathbf{x} = 0$. For sliding to occur, the condition given by (A.27) must be true for some $0 < \lambda < 1$. In other words, the convex hull of the solution of the differential inclusion must have a vector that belongs to the surface. Figure A.8 shows a situation where there is sliding. Figure A.9 now shows a situation where sliding will not occur.

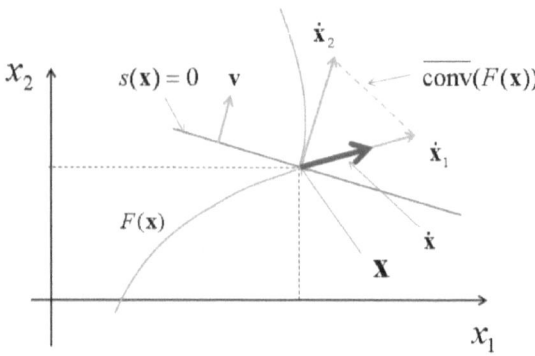

Figure A.9: Example of a situation where, on reaching the surface $\mathbf{v}^t\mathbf{x} = 0$, there is no sliding.

We can observe that, at the point $\mathbf{x}$, the convex hull of $F(\mathbf{x})$ does not contain any vector that is tangential to the surface $\mathbf{v}^t\mathbf{x} = 0$. For sliding to occur, we must have λ_f of (A.30) between 0 and 1. This implies that the states for which there is sliding must satisfy

$$0 < \frac{1}{2} - \frac{\mathbf{v}^t\mathbf{A}\mathbf{x}}{2\mathbf{v}^t\mathbf{b}} < 1 \qquad\qquad (\text{A.32})$$

which leads to

$$1 \geqslant \frac{\mathbf{v}^t\mathbf{A}\mathbf{x}}{\mathbf{v}^t\mathbf{b}} \geqslant -1 \qquad\qquad (\text{A.33})$$

The condition (A.33) establishes a set of states for which sliding occurs. This condition is equivalent to stating that the value of λ_f in (A.30) must be between 0 and 1. Hence, it is possible to define two conditions for sliding:

$$C_1 : \frac{1}{2} - \frac{\mathbf{v}^t\mathbf{A}\mathbf{x}}{2\mathbf{v}^t\mathbf{b}} > 0 \Rightarrow \begin{cases} \mathbf{v}^t\mathbf{A}\mathbf{x} < \mathbf{v}^t\mathbf{b}, & \mathbf{v}^t\mathbf{b} > 0 \\ \mathbf{v}^t\mathbf{A}\mathbf{x} > \mathbf{v}^t\mathbf{b}, & \mathbf{v}^t\mathbf{b} < 0 \end{cases} \qquad (\text{A.34})$$

$$C_2 : \frac{1}{2} - \frac{\mathbf{v}^t \mathbf{Ax}}{2\mathbf{v}^t \mathbf{b}} < 1 \Rightarrow \begin{cases} \mathbf{v}^t \mathbf{Ax} > -\mathbf{v}^t \mathbf{b}, & \mathbf{v}^t \mathbf{b} > 0 \\ \mathbf{v}^t \mathbf{Ax} < -\mathbf{v}^t \mathbf{b}, & \mathbf{v}^t \mathbf{b} < 0 \end{cases} \quad \text{(A.35)}$$

For sliding to occur, both conditions C_1 and C_2 must be met at the same time. Figure A.10 illustrates these conditions geometrically.

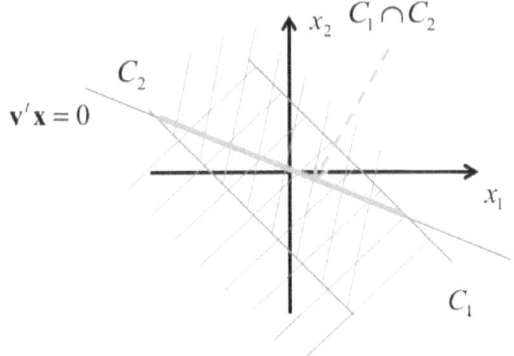

Figure A.10: Conditions for sliding on the surface $\mathbf{v}^t \mathbf{x} = 0$.

In the figure the central line indicates the surface given by $\mathbf{v}^t \mathbf{x} = 0$. The hashed regions indicate the states that satisfy each condition C_1 and C_2. Finally, the intersection is indicated by the tick central line.

An interesting situation occurs when the states belonging to the sliding surface are in the direction of one of the eigenvalues of $\mathbf{A}$. In this case the following situation occurs

$$\mathbf{Ax} = c\mathbf{x} \quad \text{(A.36)}$$

In this situation, the term present in the sliding condition (A.33) becomes

$$\frac{\mathbf{v}^t \mathbf{Ax}}{\mathbf{v}^t \mathbf{b}} = \frac{c\mathbf{v}^t \mathbf{x}}{\mathbf{v}^t \mathbf{b}} \quad \text{(A.37)}$$

As the state $\mathbf{x}$ is itself on the sliding surface ($\mathbf{v}^t\mathbf{x} = 0$) we have that $\frac{\mathbf{v}^t\mathbf{Ax}}{\mathbf{v}^t\mathbf{b}} = 0$. This satisfies (A.33) for all $\mathbf{x}$ inside the surface. In other words, the sliding surface becomes unbounded.

A.5.4 Exiting States

As emphasised before, once on the sliding surface, states will slide governed by the dynamics given by (A.31). It is possible that this dynamics will take the states in the direction of one of the limits of the sliding surface. If this is the case, the states will reach a point where there is no more sliding and they will exit the sliding surface. To understand how this happens, let us see what happens to a vector that has a tangential trajectory to some limiting state in the sliding surface. At the bounds (limits), the vector that is tangential to the trajectory is given by the derivative of the states as given by (A.31). Multiplying (A.31) by $\mathbf{v}^t$ and proceeding with the algebra we have

$$\mathbf{v}^t\dot{\mathbf{x}} = \mathbf{v}^t\mathbf{Ax} - \mathbf{v}^t\frac{\mathbf{bv}^t}{\mathbf{v}^t\mathbf{b}}\mathbf{Ax} \qquad (A.38)$$

According to the condition C_1 (or C_2 depending on the desired limit), we have that (at the boundary)

$$\mathbf{v}^t\mathbf{Ax} = \mathbf{v}^t\mathbf{b} \qquad (A.39)$$

Substituting this result in (A.38) we obtain $\mathbf{v}^t\dot{\mathbf{x}} = \mathbf{v}^t\mathbf{b} - \frac{\mathbf{v}^t\mathbf{b}}{\mathbf{v}^t\mathbf{b}}\mathbf{v}^t\mathbf{b}$, which leads to $\mathbf{v}^t\dot{\mathbf{x}} = 0$. This result is very interesting because it shows that the exit must always occur tangentially to the surface. One important question is: Once out of the sliding surface, along which of the semi-planes will the state continue? To answer this question we can use the fact that the exit is always tangential to the surface and write the semi-plane equations as

$$\dot{\mathbf{x}} = \mathbf{Ax} + \mathbf{b}$$
$$\dot{\mathbf{x}} = \mathbf{Ax} - \mathbf{b} \qquad (A.40)$$

We can have the state exiting at either $\mathbf{v}^t\mathbf{Ax} = \mathbf{v}^t\mathbf{b}$ or $\mathbf{v}^t\mathbf{Ax} = -\mathbf{v}^t\mathbf{b}$, representing both limits. If the state leaves via

$\mathbf{v}^t \mathbf{A} \mathbf{x} = \mathbf{v}^t \mathbf{b}$ then we multiply the expressions in (A.40) by $\mathbf{v}^t$ and, applying (A.39), we obtain

$$\mathbf{v}^t \dot{\mathbf{x}} = \mathbf{v}^t \mathbf{A} \mathbf{x} + \mathbf{v}^t \mathbf{b} = 2\mathbf{v}^t \mathbf{b}$$
$$\mathbf{v}^t \dot{\mathbf{x}} = \mathbf{v}^t \mathbf{A} \mathbf{x} - \mathbf{v}^t \mathbf{b} = 0 \tag{A.41}$$

As the exit must be a tangent ($\mathbf{v}^t \dot{\mathbf{x}} = 0$), the trajectory will continue via the semi-plane given by

$$\dot{\mathbf{x}} = \mathbf{A} \mathbf{x} - \mathbf{b} \tag{A.42}$$

Next, some examples that illustrate the solution given by (A.31) for the sliding surface as well as cases of the analysis of the previous sections will be presented.

A.5.5   Examples

Figures A.11 to A.15 shows some examples of systems of the form (A.2). In the figures, the vector field is indicated by the arrows that fill the plot. The lines that have arrows attached to them are the surfaces $\mathbf{v}^t \mathbf{x} = 0$. There is sliding where the arrows are darker. In all of the figures, the trajectory shown has initial conditions given in the label. One can see that where there is sliding, the convex hull formed by the vectors (arrows) always touches the line $\mathbf{v}^t \mathbf{x} = 0$, while outside the sliding surface the vectors point to the same side (their convex hulls do not contain the line $\mathbf{v}^t \mathbf{x} = 0$).

Figure A.11 illustrates a case where the sliding surface is infinite. One of the eigenvalues of $\mathbf{A}$ is $\mathbf{x} = [0 \quad -1]^t$, which is perpendicular to $\mathbf{x} = [-3 \quad 0]^t$. In other words, all vectors $\mathbf{x}$ present at the surface are proportional to one of the eigenvectors of $\mathbf{A}$.

Figure A.12 is a typical case where the sliding surface is limited and the vector field becomes stable at the surface.

The possibility for limit cycles is demonstrated in figure A.13. This is one of the cases where the limit cycle does not contain the sliding surface (which is stable).

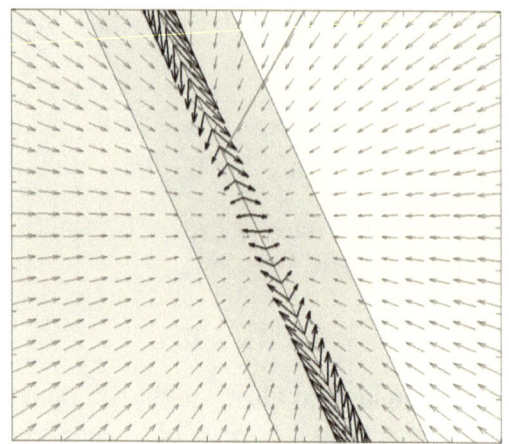

Figure A.11: Plot of the vector field for the system $\dot{\mathbf{x}} = \begin{bmatrix} -1 & 0 \\ 0 & -1 \end{bmatrix} \mathbf{x} + \begin{bmatrix} 1 \\ 0.4 \end{bmatrix} \text{SIGN} \left(\begin{bmatrix} -3 & 0 \end{bmatrix} \mathbf{x} \right)$ with initial condition $\mathbf{x}_0 = \begin{bmatrix} 2 & 10 \end{bmatrix}^t$.

Figure A.14 shows a very impressive case. There are two twin limit cycles, both containing the sliding surface. This is a remarkable example because it shows that limit cycles that contain the sliding surface (or a piece of it, as in this case) are reached in finite time! This happens because, as part of the limit cycle trajectory has the sliding surface on it, once the sliding surface has been reached, the states will go back to it, exiting every time at the same place (and entering also through the same place), thus repeating the cycle forever.

Finally, figure A.15 illustrates a case where is no sliding surface at all. In this case, since $\mathbf{v}^t \mathbf{b} = 0$, the condition (A.33) can never be satisified.

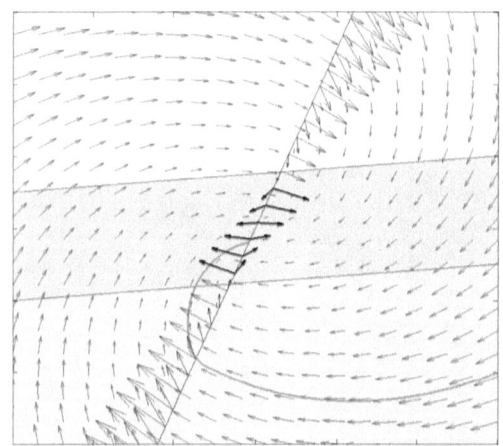

Figure A.12: Plot of the vector field for the system $\dot{\mathbf{x}} = \begin{bmatrix} -1 & 2 \\ -2 & -1 \end{bmatrix} \mathbf{x} + \begin{bmatrix} -1 \\ 0.4 \end{bmatrix}$ SIGN $\left(\begin{bmatrix} 9 & 0 \end{bmatrix} \mathbf{x} \right)$ with initial condition $\mathbf{x}_0 = \begin{bmatrix} 15 & -10 \end{bmatrix}^t$.

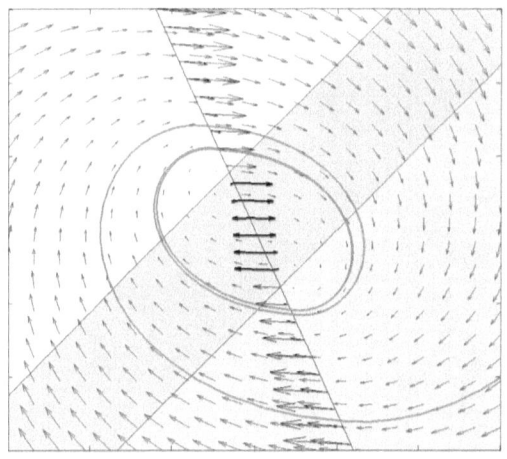

Figure A.13: Plot of the vector field for the system $\dot{\mathbf{x}} = \begin{bmatrix} -1 & 2 \\ -2 & -1 \end{bmatrix} \mathbf{x} + \begin{bmatrix} 1 \\ 0.4 \end{bmatrix}$ SIGN $\left(\begin{bmatrix} 9 & 0 \end{bmatrix} \mathbf{x} \right)$ with initial condition $\mathbf{x}_0 = \begin{bmatrix} 15 & -10 \end{bmatrix}^t$.

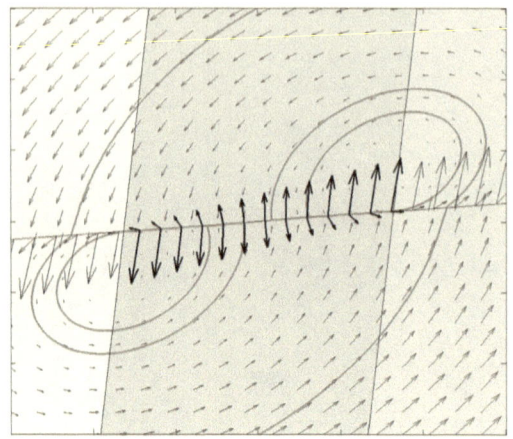

Figure A.14: Plot of the vector field for the system $\dot{\mathbf{x}} = \begin{bmatrix} 0.7 & -1.9 \\ 1.7 & -0.3 \end{bmatrix} \mathbf{x} + \begin{bmatrix} -0.1 \\ 1.8 \end{bmatrix} \text{SIGN}\left(\begin{bmatrix} 0.6 & -12.4 \end{bmatrix} \mathbf{x}\right)$ with initial condition $\mathbf{x}_0 = \begin{bmatrix} 0 & 15 \end{bmatrix}^t$ and $\mathbf{x}_0 = \begin{bmatrix} 0 & -15 \end{bmatrix}^t$.

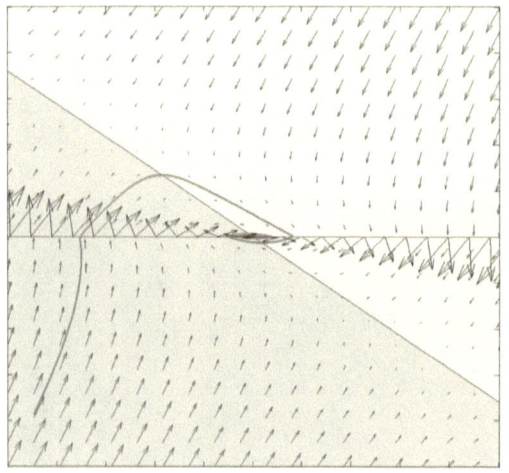

Figure A.15: Plot of the vector field for the system $\dot{\mathbf{x}} = \begin{bmatrix} -0.3 & -1 \\ -0.8 & -1.1 \end{bmatrix} \mathbf{x} + \begin{bmatrix} 0 \\ 1 \end{bmatrix} \text{SIGN}\left(\begin{bmatrix} 2.9 & 0 \end{bmatrix} \mathbf{x}\right)$ with initial condition $\mathbf{x}_0 = \begin{bmatrix} -10 & -10 \end{bmatrix}^t$.

Bibliography

[1] S. O. Haykin, *Adaptive Filter Theory (4th Edition)*. Prentice Hall, 4 ed., Sept. 2001.

[2] A. Papoulis, *Probability, Random Variables, and Stochastic Processes*. Mc-Graw Hill, 1984.

[3] A. F. Filippov, *Differential Equations with Discontinuous Righthand Sides: Control Systems*. Springer, 1 ed., Sept. 1988.

[4] V. I. Utkin, *Sliding Mode Control: Mathematical Tools, Design and Applications*, vol. 1932 of *Lecture Notes in Mathematics*. Berlin, Heidelberg: Springer Berlin / Heidelberg, 2008.

[5] N. S. Nise, *Control Systems Engineering*. Wiley, 5 ed., Dec. 2007.

[6] K. Ogata, *Modern Control Engineering (4th Edition)*. Prentice Hall, Nov. 2001.

[7] R. C. Dorf and R. H. Bishop, *Modern Control Systems (11th Edition) (Pie)*. Prentice Hall, Aug. 2007.

[8] S. Haykin, *Neural Networks: A Comprehensive Foundation (2nd Edition)*. Prentice Hall, 2 ed., July 1998.

[9] S. Haykin, *Communications Systems*. Wiley, 4th ed., May 2000.

[10] S. Haykin and B. Van Veen, *Signals and Systems, 2nd Edition*. Wiley, 2 ed., Oct. 2002.

[11] G. Strang, *Linear Algebra and Its Applications*. Brooks Cole, 4th ed., July 2005.

[12] G. Strang, *Computational Science and Engineering*. Wellesley-Cambridge Press, 1 ed., Nov. 2007.

[13] E. Parzen, "On the estimation of a probability density function and mode," *Annals of Mathematical Statistics*, vol. 33, pp. 1065–1076, 1962.

[14] J. C. Principe, *Information Theoretic Learning*. New York, NY: Springer New York, 2010.

[15] W. Liu, J. C. Principe, and S. Haykin, *Kernel Adaptive Filtering: A Comprehensive Introduction*. Wiley Publishing, 1st ed., 2010.

[16] J. T. George and R. L. Finney, *Calculus and Analytic Geometry (9th Edition)*. Addison Wesley Publishing Company, Jan. 1996.

[17] J. H. Hubbard and B. B. Hubbard, *Vector Calculus, Linear Algebra, and Differential Forms: A Unified Approach*. Prentice Hall, 2nd ed., Sept. 2001.

[18] B. van Brunt, *The calculus of variations*. Universitext, New York: Springer-Verlag, 2004.

www.ingramcontent.com/pod-product-compliance
Lightning Source LLC
Chambersburg PA
CBHW032002170526
45157CB00002B/507